CLA
LOCOM
from 1850 to th

CW00457350

...day

DENIS GRIFFITHS

CLASSIC LOCOMOTIVES
from 1850 to the present day

DENIS GRIFFITHS

Patrick Stephens
Wellingborough, Northamptonshire

First published 1988

British Library Cataloguing in Publication Data

Griffiths, Denis, *1943-*
Classic locomotives.
1. Locomotives—Great Britain
I. Title
625.2'6'0941   TJ605

ISBN 0-85059-933-4

Dedication

For
SARAH
my daughter
with love

Title page *Broad Gauge replica* Iron Duke *at Didcot.*
Cover illustrations
Front *'West Country' 'Pacific' No 34092* City of Wells *on the Worth
Valley Railway in May 1987.*
Back *'Black Five' No 45110 at Manchester Victoria, 4 August 1968.*

*Patrick Stephens Limited is part of the
Thorsons Publishing Group, Wellingborough,
Northamptonshire, NN8 2RQ, England.*

Printed in Great Britain by Richard Clay Limited,
Chichester, Sussex

1 3 5 7 9 10 8 6 4 2

CONTENTS

INTRODUCTION

What is a classic locomotive? The locomotive part is easy but arguments will abound as to which locomotives can be reasonably described as 'classic'. Personal preference will certainly influence the individual's opinion and that preference might well result from experience or familiarity rather than reason. The dictionary defines classic as being 'a work of recognized excellence'. But who decides which locomotives are excellent? The simple answer to that question is the reader of this book. We all have our own ideas as to what constitutes excellence and we all have our own preferences or prejudices which will colour our decision. However, our choice will be made on the basis that there is a measure of excellence which can be attributed to that particular locomotive class.

It is natural that the Great Western enthusiast would tend to favour Swindon products whilst the LMS devotee might be blinkered to all but Stanier machines. Even within those narrow confines there is scope for a selection procedure as to which of the locomotives could be considered classic rather than just being very good. The problem then is one of deciding which criteria are to be employed in making the judgement. Many locomotives introduced new features which then became standard, but that does not in itself make them classic machines. Longevity in service is a good criterion, as only useful engines were retained and developed over the years; but is that alone justification? Some classes were exceptionally fast, others were powerful but many were simply hard-working and reliable. There are locomotives which can be considered as attractive, even handsome, but beauty is in the eye of the beholder.

These are all attributes which can be used, singly or in combination, to confer the title 'classic'. There is, however, something else—an indefinable quality which can set the classics apart from the rest, and only the individual can say what that is as far as he is concerned. Even then personal preferences will intervene, but that is conditioning and experience. In effect, there is no real definition of a classic locomotive, so we are all correct; for who is to provide a definite argument against?

There are fifty locomotive types described in this book and they could all, as far as the author is concerned, be described as classic. The reader may not agree with all fifty, but the selection is so varied that few would totally disagree. No attempt was made to please everybody, but in any list of fifty it would be almost impossible not to include some favourite of every British railway enthusiast. Had there only been ten in the list, the task would have been more difficult.

Express passenger locomotives are always considered as being something special, and they are well represented here, but what about the less glamorous shunting and freight engines? Generally neglected by the enthusiast, they were the work-horses of the railway system, performing under the most difficult of conditions but with high availability. Who would deny them their place in this book? Little engines like the L & NWR 'Precedents' and GNR Stirling 'singles' earn inclusion because of their valiant exploits in the races to Scotland so many years ago, whilst the GWR Dean single-wheelers are just too beautiful to be ignored. Diesels may not be favourites with steam enthusiasts but few steamers received the adulation bestowed upon the 'Westerns', 'Deltics' and Class '40s' during their declining months.

In order to avoid any intimation of preference, the locomotives have been arranged in chronological order. Such a sequence also allows comparisons to be made between the products of different railway companies and their engineers. Churchward's outstanding design achievements at the beginning of the twentieth century are well-known, as are the advances under Gresley and Stanier during the 1930s, and although all designs should be considered in the light of a railway's requirements, few people would deny that these engineers left their mark on the British locomotive scene.

If nothing else, the reader may use this book as a valuable and compact reference source, the technical information being detailed and the locomotive 'biographies' comprehensive. Hopefully, it will also serve another purpose. This volume details one enthusiast's collection of classic locomotives, and with luck it will stimulate the reader to consider his own list, not just from a narrow preferred railway view but from the wider British context. The experience will be worthwhile and rewarding.

GWR BROAD GAUGE 'IRON DUKE' CLASS 2+2-2-2

Designer Daniel Gooch. **Introduced** 1847. **Driving wheel diameter** 8 ft 0 in. **Carrying wheel diameter** 4 ft 6 in. **Cylinders** Two (inside) 18 in diameter × 24 in stroke. **Valve gear** Gooch. **Boiler** (as fitted to the initial batch of 1847)—*Pressure* 100 psi (later 115 psi). *Grate area* 21.66 sq ft. **Heating surfaces**—*Tubes* 1,797 sq ft. *Firebox* 147.9 sq ft. **Weight** Loco 35 tons 10 cwt, tender 21 tons. **Number built** 30 (including rebuilt *Great Western* of 1846).

Whilst the Great Western Railway was locked in battle with other companies over the matter of a gauge for the railways of Britain, there was considerable competition as far as speed and hauling power was concerned. Early in 1846, Gooch was given permission to construct a powerful express engine and barely 13 weeks later *Great Western* emerged from Swindon works. It was a 2-2-2 with 8 ft diameter driving wheels and a large 'haycock' boiler. The directors' confidence in the young Daniel Gooch was rewarded when the locomotive showed itself capable of hauling 100-ton loads at average speeds in excess of 60 mph over long distances.

However, a major fault soon became evident when the leading axle broke whilst running at speed near Shrivenham; the ingenious Gooch rectified matters by extending the frames and fitting a second set of leading wheels to make a 2+2-2-2 arrangement. That set the pattern for large-scale construction of the 'Iron Duke' Class, which also received larger, round-topped boilers. Constructed from 1847 to 1855, with the final batch being built by Rothwell & Co rather than at Swindon, these Gooch singles proved themselves to be superior to any contemporary locomotive on any railway.

Their performance on the main line to Bristol was legendary, and for most people they epitomized the broad gauge. Not all were identical, there being slight differences in wheelbase and boiler dimensions, but visually they looked the same and performed the same. Gooch provided no cabs on any of his locomotives but the massive firebox of the singles provided more protection than was offered by the rudimentary cabs fitted to some narrow gauge engines. In later years, simple cabs were added but their introduction lagged behind that on other railways; GWR footplate crews had to be hardy.

Operations were not without incident. The boiler of *Perseus*

Timour, *a broad gauge rebuild of 1873 in its final form.*

exploded whilst at Westbourne Park shed on November 8 1862 whilst on October 1 1852 a special train hauled by *Lord of the Isles* had contrived to run into a preceding train whilst taking the directors to Birmingham for the celebration of the opening of the broad gauge route to that city.

By 1870, some of the locomotives were showing their age and renewal was required. It was also evident that the days of the broad gauge were limited, but a completely new design would have been justified. However, the new engines were almost identical to the original Gooch design, an indication of the durability and soundness of his engineering skill. Nonetheless, they were effectively new locomotives as few of the original parts were incorporated. The final trio did not appear until 1888, only four years before the final elimination of the broad gauge. It was these later locomotives, with their cabs and flanged driving wheels, which became well-known through the work of late nineteenth-century railway photographers.

One of the original locomotives, *Lord of the Isles*, was withdrawn in 1884 for preservation, and was exhibited in Britain and Chicago during the 1890s. A senseless act of vandalism on the part of G. J. Churchward resulted in this beautiful machine being scrapped in 1906, but fortunately a replica of *Iron Duke* was later constructed using a War Department austerity 0-6-0 as its basis.

GNR 8 FT 'SINGLE' 4-2-2

Designer Patrick Stirling. **Introduced** 1870 (1895). **Driving wheel diameter** 8 ft 1 in (8 ft 1.5 in). **Leading wheel diameter** 3 ft 11 in (3 ft 11.5 in). **Trailing wheel diameter** 4 ft 1 in (4 ft 7.5 in). **Cylinders** Two (outside) 18 in diameter × 28 in stroke (19 in × 28 in). **Valve gear** Stephenson. **Boiler**—*Pressure* 140 psi (170 psi). *Grate area* 15.8 sq ft (20 sq ft). **Heating surfaces**—*Tubes* 875.5 sq ft (910 sq ft). *Firebox* 92.5 sq ft (121.7 sq ft). **Tractive effort** 11,130 lbs. **Weight** Loco 38 tons 10 cwt, tender 30 tons. **Number built** 53. (The first figures apply to No 1, the first of the 8 ft singles, while those in brackets apply to the final locomotive, No 1008)

The Great Northern Railway from King's Cross to York formed the first leg of the East Coast Route between London and Scotland. Fast, powerful locomotives were required to operate the passenger services and in 1868, two years after his arrival, Stirling appears to have formed the opinion that the machines available were not up to the task. After borrowing an outside-cylindered 7 ft single from the Great Eastern Railway, he settled his mind on producing a superior locomotive. His choice of 8 ft diameter driving wheels was made on the basis of better adhesion with the rail; larger diameter wheels also reduced the local stress intensity on the rail by spreading the load over a larger area of contact.

The prototype, No 1, emerged from Doncaster works in April 1870. By contemporary standards it was graceful and attractive, with the valve gear hidden between the frames, but inside cylinders, which were preferred in Victorian times, could not be contemplated for two reasons, both due to the large driving wheels. Side force on the wheel flanges would have caused excessive leverage on the cranks, and a boiler centre-line height of nearly 8 ft would have been required to clear the cranks. The final outcome of a curved transition from the smokebox to the outside cylinders was pleasing to the eye, whilst when running the flashing rods gave an impression of speed.

Early defects with No 1 were soon overcome and other examples followed, but very slowly, averaging two a year until 1877. Modifications meant that no members of the class were ever identical. During the 1880s, earlier examples were rebuilt and reboilered, but the same basic shape prevailed. The final five of the class did not appear until 1895, the year in which Stirling died, still in harness.

Despite the differences, the class as a whole was a considerable success, and Stirling's 8 ft single-wheelers were the mainstay of the Great Northern's express passenger trains, being capable of sustained speeds in excess of 80 mph on normal trains during the 1890s when working with the higher boiler pressure. This performance has to be judged in the light of prevailing conditions and not by the standards of later years, but, for the time, it was good and the locomotives economical.

The 'Races to the North' established the reputation of these machines. The 1888 rivalry between the East and West Coast Routes for the shortest journey times between London and Scotland soon developed into an all-out race. The 'singles' operated the service between London and York with an engine change at Grantham, and start-to-stop average speeds of 55 mph were frequently exceeded with trains of 130 tons or more. For lighter trains, 60 mph was not uncommon. The 1895 'races' to Aberdeen produced even more remarkable results.

Such performances were certainly classic, establishing the reputation of Stirling and his 8 ft 'singles'. It was, however, the swan-song of the single-wheeler; as trains became heavier, the 'singles' were less effective and gradually gave way to the Ivatt 'Atlantics', classic locomotives in their own right (see page 32).

No 1007, one of the last batch of Stirling 8 ft 'singles' (National Railway Museum, York).

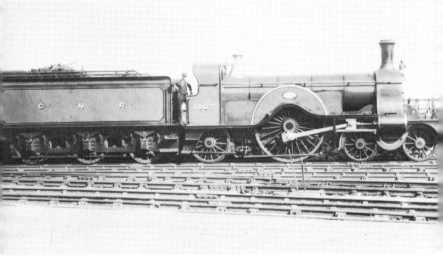

LB & SCR 'TERRIER' 0-6-0 TANK

Designer William Stroudley. **Introduced** 1872. **Power classification** OP. **Driving wheel diameter** 4 ft 0 in. **Cylinders** Two (inside) 13 in diameter × 20 in stroke (some later versions had 12 in and 14 in diameters). **Valve gear** Stephenson. **Boiler**—*Pressure* 140 psi (150 psi). *Grate area* 10.3 sq ft (10 sq ft). **Heating surfaces**—*Tubes* 473 sq ft (433.2 sq ft). *Firebox* 55 sq ft (55.6 sq ft). **Tractive effort** 8,890 lbs for 12 in diameter cylinders. **Weight** 26 tons 17 cwt (28 tons 10 cwt). **Number built** 50. (The figures in brackets apply to the final years in British Railways service)

It is difficult to believe that a class of locomotives designed in 1872 would survive in normal railway service for over ninety years. It is even more amazing to realize that some members of the class still operate today in revenue-earning service, albeit on preserved railways. The design was originally intended to furnish motive power for operating the South London suburban services of the London, Brighton and South Coast Railway which were then just beginning to expand. Six locomotives were originally constructed at Brighton and they astonished passengers and railway staff alike with their performance and versatility. The nickname 'Terrier' was soon applied and it stuck (although an alternative nickname, 'Rooters', found initial favour with the drivers).

The locomotives' names appeared in large letters across the side tanks, and were taken from London areas such as Poplar, Wapping and Fenchurch. A further batch was constructed in 1874 and more were built in the years up to 1880. So successful were they at operating the suburban services that traffic increased beyond their means and more powerful traction was called for. Natural expansion of the suburban areas was mainly responsible, but the efficient and reliable services operated by the 'Terriers' did much to popularize rail commuting from London's southern suburbs.

One member of the class, No 40 *Brighton*, went overseas to the Paris Exhibition of 1878 and gained a gold medal. Demonstration running in France showed the economy of the little engines as well as the efficiency of the Westinghouse brake system which was fitted to all new members of the class after the initial six.

Removal from the main commuter services did not mean disposal, as there was valuable work to be performed on branch

Stroudley 'Terrier' No 32636 in BR days (D. K. Jones Collection).

lines. The 'Terriers' were also moved to duties as station pilots, an essential but unglamorous task during the steam era. In 1899 it was decided that no more than 15 of the class would be required for future duties and the decision was made to scrap the remainder. Fortunately, R. J. Billinton, then Locomotive Superintendent of the LB & SCR, realized that they had a resale value for use on light railways. Some did not find buyers and were eventually cut up, but several were purchased for operating services on the Isle of Wight Railway and a number were used by Pauling & Co, the railway contractors, for constructing lines elsewhere; several of these were ultimately converted to tram engines for service in South America. Smaller railways in Britain bought several as did the Admiralty for use in its dockyards; by that route, No 37 *Southdown* operated as far north as Inverness.

Those that remained in LB & SCR service continued to perform sterling work and many were subsequently taken over by the Southern Railway and then British Railways. Upon nationalization, a number of those originally sold rejoined their sisters under the common owner.

Two 'Terriers' still operate on the preserved Isle of Wight Railway whilst others can be seen pulling trains on the Bluebell Railway and the Kent & East Sussex Railway. Two form part of the National Collection, whilst one, No 54 *Waddon*, has even emigrated to Canada.

L & NWR 'PRECEDENT' CLASS 2-4-0

Designer F. W. Webb. **Introduced** 1874. **Driving wheel diameter** 6 ft 6 in. **Leading wheel diameter** 3 ft 9 in. **Cylinders** Two (inside) 17 in diameter × 24 in stroke. **Valve gear** Allan. **Boiler**—*Pressure* 150 psi. *Grate area* 17.1 sq ft. **Heating surfaces**—*Tubes* 380 sq ft. *Firebox* 94.6 sq ft. **Tractive effort** 10,920 lbs. **Weight** Loco 35 tons 12 cwt, tender 25 tons. **Number built** 166.

To misuse a common expression, some locomotives are built great whilst others have greatness thrust upon them. The latter would appear to be the case with the London & North Western's 'Precedent' Class, frequently referred to as 'Jumbos'. Early performances were not spectacular but, when faced with hauling prodigious loads north from Euston and Crewe during the 1895 races to Aberdeen, they more than proved themselves to be classic.

L & NWR 'Precedent' Class No 790 Hardwicke *at Dinting.*

John Ramsbottom introduced the 'Newton' Class of 2-4-0 locomotives to the L & NWR in 1866 and production continued at a steady rate until 1876. Webb, who had been Ramsbottom's Chief Draughtsman, played a considerable part in the design and continued production of these engines until 1873, and when the first of the 'Precedents' appeared a year later and they were very much in the same mould as the 'Newtons' but with a larger boiler. In 1887, the 'Improved Precedent' or 'Large Jumbo' was introduced, of the same basic design but with the boiler pressure raised to 150 psi and stronger frames.

Much confusion resulted from the L & NWR's rebuilding and numbering policy, if there was actually such a thing as a policy. When old 'Newtons' were scrapped they were replaced by 'Improved Precedents' using the same number-plates and name-plates. As these usually included a building date it caused the incorrect conclusion to be drawn that the engines were rebuilds and not of a completely new construction. A further confusion arose when a version identical but for its 6 ft driving wheels was constructed between 1889 and 1896, being known as the 'Whitworth' Class. Both types came under the nickname 'Jumbo' but it was only those locomotives with 6 ft 6 in driving wheels which were officially classed as 'Precedents'.

During their first ten years of service the 'Precedents' performed well, but the train schedules were not fast and did not call for high-speed running. Webb's enthusiasm for compounding pushed the class away from the more glamorous trains of the late 1880s as the three-cylinder compound engines were put on almost anything which could be classed as an express. It soon became apparent, however, that they were not up to the job of fast running and 'Jumbos' of both the 6 ft 6 in and 6 ft versions soon took over the main expresses. In some cases, double-heading was employed for the 1 in 75 gradients over Shap, but still the reputation was not established.

Events during the summer of 1895 changed all that and firmly established the 'Precedents' in the league of the all-time greats. The night sleeper service to Aberdeen, run in conjunction with the Caledonian Railway, was accelerated in response to direct competition being offered by the East Coast Route. 'Precedents' were allocated to the sections between Euston and Crewe as well as Crewe to Carlisle; they were not specially modified or tuned, simply set aside for that service. No 790 *Hardwicke* was assigned to the northern section and consistently hauled loads of 12

Webb 'Precedent' Hardwicke *in company with a Webb 0-6-2 'Coal Tank' at Dinting.*

six-wheeled coaches at an average speed in excess of 48 mph. By the end of July, the schedules had quickened and average speeds of nearly 53 mph were achieved for the run over Shap. Locomotives on the southern section were performing equally well. Towards the end of the six-week racing period, *Hardwicke* achieved an average speed of 63.1 mph for the 141 miles between Crewe and Carlisle, whilst No 1213 *The Queen* managed an average of 62.7 mph—truly spectacular running.

One of the final runs saw *Hardwicke* average 65.3 mph for the climb over Shap, the journey from Crewe to Carlisle being accomplished at an average speed of 67.2 mph, and that achievement was not beaten for over forty years. It might be argued that the loads were light, and so they were by later standards, but the performances on a train weight to locomotive weight basis match the best achieved by subsequent more powerful locomotives.

It is fitting that *Hardwicke* has been preserved as the representative of this truly remarkable class of locomotives.

FESTINIOG RAILWAY 'DOUBLE FAIRLIE' 0-4-4-0 TANK

Designer Robert Fairlie/G. P. Spooner. **Introduced** 1879 (1979). **Driving wheel diameter** 2 ft 8 in. **Cylinders** Four (two outside on each bogie) 9 in diameter × 14 in stroke. **Valve gear** Stephenson. **Boiler**—*Pressure* 160 psi. *Grate area* 12.1 sq ft (14.32 sq ft). **Heating surfaces**—*Tubes* 817.2 sq ft (462 sq ft). *Firebox* 69.9 sq ft (43.5 sq ft). *Superheater* 169 sq ft (Earl of Merioneth only). **Tractive effort** 6,059 lbs (9,639 lbs). **Weight** 24 tons (32 tons, working). (The first figures relate to *Merddin Emrys* as originally constructed. Figures in brackets refer to *Earl of Merioneth* where different)

The narrow gauge Festiniog Railway in North Wales operates on some very severe continuous gradients between Porthmadog and Blaenau Ffestiniog. Curves on the 1 ft 11.5 in gauge track are also very sharp, thus limiting the size of locomotive which can be employed. Steam operations commenced in 1863, but booming slate and passenger traffic resulted in problems when the original locomotives were unable to haul the increasingly heavier trains. Limitations imposed by the line itself seemed to present an insurmountable obstacle when standard engineering practice dictated a larger engine to produce more power.

Robert Fairlie provided the solution with his double-ended articulated locomotive already operating in standard gauge form. Suitably modified for the Festiniog gauge, it would essentially be the same as two locomotives placed back-to-back. The engine units are actually built into the bogies which are connected to the frames by means of pivots, the steam and exhaust being transmitted by flexible connections. The boiler unit runs the whole length of the firebox, and, contrary to popular misconception, is actually a single boiler with two separate fireboxes. The driver and fireman are positioned on opposite sides of the boiler and the coal is fed into the fireboxes from the side. When these engines were coal burning, storage of sufficient fuel caused some problems, but subsequent conversion to oil firing has eliminated that difficulty.

The first Fairlie patent locomotive for the Festiniog was *Little Wonder*, built in 1869 by the Fairlie Engine & Steam Carriage Co. A second, *James Spooner*, followed three years later and both operated the line with a reasonable degree of success. So happy was Robert Fairlie with the outcome that he allowed the Festiniog to construct locomotives to his design without payment of royalties. The company constructed two updated versions in 1879,

Merddin Emrys and *Livingston Thompson*, at its own Boston Lodge works. The former locomotive is still in service and, apart from the cab, looks very much the same as it did when first built.

The railway itself underwent a period of severe decline and was eventually closed before being rescued by devoted preservationists. Growing tourist traffic in the late 1960s necessitated a good stable of locomotives but the boilers on the two Boston Lodge built 'Double Fairlies' were becoming life-expired. Two replacement superheated boilers were ordered from the Hunslet Engine Co but only that of *Merddin Emrys* was replaced. A bold decision was taken not to rebuild the other engine but to construct an entirely new 'Double Fairlie' at Boston Lodge. This would allow some of the limitations of the original design to be overcome in the light of more recent technology.

The new machine, named *Earl of Merioneth*, made use of some parts from the now redundant *Livingston Thompson* but it was made more bulky than the original design to accommodate additional fuel and water. The frame, however, remained the same in order to comply with the strict Festiniog loading gauge. That twentieth-century Fairlie took almost ten years to complete but the results were well worth the waiting. For an essentially volunteer railway to construct its own locomotive still stands as one of railway preservation's finest achievements and must have encouraged later standard gauge schemes.

The 'Double Fairlie' concept was remarkable, but to have two such locomotives at work today, one over 100 years old and the other of such recent construction, is outstanding.

Earl of Merioneth, *the 1979 'Double Fairlie'.*

GWR 'DEAN GOODS' OR '2301' CLASS 0-6-0

Designer William Dean. **Introduced** 1883. **Power classification** 2MT. **Driving wheel diameter** 5 ft 2 in. **Cylinders** Two (inside) 17 in diameter × 24 in stroke (in 1908 17.5 in diameter cylinders became standard). **Valve gear** Stephenson. **Boiler** —*Pressure* 140 psi (180 psi). *Grate area* 16.4 sq ft (15.45 sq ft). **Heating surfaces** —*Tubes* 1,079 sq ft (961 sq ft). *Firebox* 113.6 sq ft (106.5 sq ft). *Superheater* 97 sq ft (where fitted). **Tractive effort** 13,313 lbs (18,140 lbs in the final form) **Weight** Loco 37 tons, tender 36 tons 15 cwt. **Number built** 280. (A large variety of boilers were fitted to the class over the years; the figures given are typical. Figures in brackets are for the superheated type later applied.)

Joseph Armstrong introduced the 0-6-0 'standard Goods' in 1867 and his extensive fleet served the GWR's freight requirements for the following 15 years. By the early 1880s, increasing traffic dictated further building but William Dean, then in command at Swindon, decided to construct an updated version incorporating inside frames. The twenty constructed in 1883 proved to be a success and further batches were ordered. The early engines had 5 ft driving wheels but this figure gradually increased over the years by 2 in through the use of progressively thicker tyres. Following the batches delivered in 1884 there was a lull in production until 1890, with steady output then continuing until 1899.

During that production period many modifications were made, particularly with regard to the boiler, but these changes were brought about by design advancements rather than inherent defects. Throughout the GWR system, the 'Dean Goods' was used extensively because of its light axle loading and high pulling power. Churchward made changes as he did with other Dean locomotives, the most drastic of which being the conversion of twenty to 2-6-2 tank engines in 1907; surplus 0-6-0 tender engines and a lack of capacity at Swindon would not allow new construction of suburban tank engines for the Birmingham area, hence the Churchward reconstructions.

Churchward's modifications to the original engines included an increase in cylinder diameter, the application of superheating and the fitting of Belpaire firebox boilers with higher steam pressure. These features improved performance and power but it is significant that Churchward never attempted to introduce an 0-6-0

Above *'Dean Goods' No 2340 in original form* (Flewellyn Collection, GWR Museum, Swindon).

Below *'Dean Goods' No 2323 with the final form of Belpaire firebox superheat boiler* (GWR Museum, Swindon).

tender engine design of his own; obviously he was satisfied with the basic Dean arrangement following its upgrading.

The First World War brought service overseas for the 'Dean Goods' when the Railway Operating Department (ROD) requisitioned 62 for duties in France and Salonika. Two of the group which served in Salonika were sold for post-war operations in that region whilst seven were destroyed. During the Second World War, the call to military duty came again despite the fact that the basic design was then over fifty years old. One hundred engines were initially requisitioned but 108 were subsequently taken. Some went to France only to be caught up in the retreat from Dunkirk; many were destroyed but some were captured and later used by the Germans. Others served in North Africa and Italy whilst several went to China after the war; at least two from the European batch ended up behind the Iron Curtain.

The general reliability and simplicity of these little engines made them most useful and their relatively light axle loading allowed them to operate where their larger brethren could not go. Service overseas tends to hide their usefulness on the GWR itself, but by the end of the 1930s they were being superseded by more modern stock. Following nationalization and the introduction of BR standard classes, withdrawal proceeded at a pace. The last survivor, No 2516, was withdrawn in 1956, some 63 years after the class was introduced, and rests in the former Wesleyan Chapel which houses the GWR Museum at Swindon.

For longevity and universal usefulness, there have been few locomotive classes which could compare with the 'Dean Goods'.

NBR 'C' CLASS (LNER 'J36' CLASS) 0-6-0

Designer Matthew Holmes. **Introduced** 1888. **Power classification** 2F. **Driving wheel diameter** 5 ft 0 in. **Cylinders** Two (inside) 18 in diameter × 26 in stroke. **Valve gear** Stephenson. **Boiler**—(figures given are for final production batches)—*Pressure* 150 psi. *Grate area* 17 sq ft. **Heating surfaces**—*Tubes* 1,130 sq ft. *Firebox* 105 sq ft. **Tractive effort** 17,900 lbs. **Weight** Loco 41 tons 19 cwt, tender 33 tons 10 cwt. **Number built** 168.

The North British Railway was the largest of the railways in Scotland with almost a monopoly of the eastern area, including the whole of Fife with its important coal reserves. With such mineral traffic it was natural that an extensive fleet of goods locomotives would be required. It was equally natural that they would be of the 0-6-0 wheel arrangement which then appeared to be standard for similar locomotives throughout Britain. In 1883, Matthew Holmes produced his first 0-6-0 goods engine but it was really a modified form of the earlier design by Dugald Drummond. However, over the next four years further engines of the class were constructed with changes introduced by Holmes.

A pure Holmes 0-6-0 tender engine design appeared in 1888 and over the following twelve years no fewer than 168 were built, mainly at the railway's own Cowlairs works although 15 were built by Neilsen & Co in 1891 and another 15 by Sharp, Stewart & Co in 1892. The early locomotives had a slightly smaller boiler than that which became standard with later batches. Holmes had obviously learned about the need for a high-capacity boiler for goods locomotives, and eventually increased the firebox heating surface to 105 sq ft, which was large for the period. On goods trains they were most effective, but their relatively high tractive effort also made them suitable for snow-plough duties during the winter months when conditions on the railway were severe. Holmes designed large wooden ploughs which were mounted directly on the front of the engine by means of bolted connections on the buffer beam. The top was connected to the chimney by a wire rope adjusted by a screw toggle, thus allowing the height of the plough to be altered.

In later years, 1913-20, William Paton Reid, Holmes' successor on the North British, rebuilt many of the class giving them larger diameter boilers and new laminated springs but they were still

North British Railway Holmes-designed 0-6-0 goods locomotive No 673 Maude.

basically as originally designed. An assortment of braking systems had been applied; 87 members of the class had steam brakes, a further 13 had steam and vacuum brakes, 48 were supplied with Westinghouse air brakes and 20 had combined Westinghouse and vacuum brake systems. Such versatility was useful on a railway where through workings from and to other systems were an essential part of operations. This was not confined to goods traffic; passenger stock was also taken whenever necessary.

During the First World War, 25 of the class were taken for work overseas in support of the military and all were subsequently returned. In recognition of their war service, these humble goods engines were given the names of notable generals and places associated with that conflict. Grouping gave them the LNER designation 'J36', but they continued with the same type of traffic and in the same area of operation for which they had been originally designed. Longevity results from usefulness and that applied to the 'J36s'. Despite the introduction of more modern locomotives for other services, it would seem that few improvements could be made to the basic Holmes design of 1888. Well over half of the class survived until the 1960s, and some until 1966. *Maude*, NBR No 673, was purchased by the Scottish Railway Preservation Society and can still be seen hauling main-line trains a century after the class was introduced.

GWR 'ACHILLES' OR '3031' CLASS 4-2-2

Designer William Dean. **Introduced** 1894. **Driving wheel diameter** 7 ft 8.5 in. **Leading wheel diameter** 4 ft 1 in. **Trailing wheel diameter** 4 ft 7 in. **Cylinders** Two (inside) 19 in diameter × 24 in stroke. **Valve gear** Stephenson. **Boiler**—*Pressure* 160 psi. *Grate area* 20.8 sq ft. **Heating surfaces**—*Tubes* 1,434 sq ft. *Firebox* 127 sq ft. **Tractive effort** 17,378 lbs. **Weight** Loco 49 tons, tender 32 tons 10 cwt. **Number built** 80.

In 1890, William Dean designed a class of 2-2-2 single-wheelers for express passenger duties, and eight of them were constructed as convertible engines; they entered service on the broad gauge but with demise of Brunel's 7 ft gauge were soon converted to standard gauge. The twenty which appeared in 1892 were built to the standard gauge. Although powerful and attractive locomotives, they were heavy at the front end and had a tendency to be unsteady whilst running at speed. That defect was forcibly brought home on September 16 1893 when No 3012 *Wigmore Castle* became derailed in Box Tunnel. Redesign to improve stability at the front resulted in an extension of the frames and the fitting of a bogie. This change produced the desired effect as far as operation was concerned and, in terms of aesthetics, its effect was even more dramatic.

The 2-2-2 form as originally built may not have been an 'ugly duckling', but the transformation certainly produced a locomotive with the grace of a swan. Little else was changed in the rebuild and it is difficult to imagine how the mere extension of the frames and the fitting of a long-wheelbase bogie could produce such a remarkable alteration. The thirty existing 2-2-2s were rebuilt in 1894 whilst others of the new form were ordered, production continuing until 1899. In terms of style, elegance and grace there have been few locomotive designs to compare with Dean's 4-2-2 '3031' Class. Brass ornamentation was appropriate and was used to adorn the dome and driving wheel splashers as well as other small parts to create the effect of opulence which personified the Great Western during that late Victorian period.

These Dean singles were not just things of beauty, they were hard-working locomotives on the main express trains to the West. They monopolized express passenger workings between Paddington and Newton Abbott and were only rarely observed away from that route before the turn of the century. The locomotives constructed as 4-2-2s differed from the earlier 2-2-2 rebuilds in that the cylinder diameter was reduced by 1 in. For one glorious year following the delivery of the final '3031' Class locomotive, *Windsor Castle*, in March 1899, there were eighty identical machines of outstanding beauty operating on Brunel's main line.

However, in the following year Churchward's views on boiler design began to gain precedence and raised Belpaire fireboxes became the order of the day; they were gradually fitted to members of the class and, although more efficient, destroyed the locomotive's graceful lines. Style gave way to the demand for efficiency.

Several subsequent rebuildings took place with different boiler designs being tried, but the frames and other engine parts remained very much the Dean original. Displaced from express services to Devon, the single-wheelers found employment on trains to the Midlands and then for local traffic in Somerset. Withdrawals began in 1908 and by the end of 1915 they had all gone. Churchward's advances in locomotive design hastened their end, but although they had relatively short operating lives these Dean single-wheelers left their mark. The GWR cared little for its past, so no example was preserved, but a non-working replica, No 3041 *The Queen*, has been constructed for Madame Tussauds and is displayed at Windsor and Eton station. It may not have the power of the originals but it still has the style and grace.

Dean single-wheeler Worcester *of the '3031' Class* (GWR Museum, Swindon).

L & SWR 'T9' CLASS 4-4-0

Designer Dugald Drummond. **Introduced** 1899. **Power classification** 3P. **Driving wheel diameter** 6 ft 7 in. **Leading wheel diameter** 3 ft 7 in. **Cylinders** Two (inside) 19 in diameter × 26 in stroke. **Valve gear** Stephenson. **Boiler** (figures given are for superheated boiler of 1922)—*Pressure* 175 psi. *Grate area* 24 sq ft. **Heating surfaces**—*Tubes* 920 sq ft. *Firebox* 142 sq ft. *Superheater* 213 sq ft. **Tractive effort** 17,670 lbs. **Weight** Loco 51 tons 16 cwt, tender 44 tons 17 cwt. **Number built** 66.

Dugald Drummond came to the London & South Western Railway in 1895 after service as Locomotive Superintendent on two Scottish railways, the North British and Caledonian. For both he designed inside-cylindered 4-4-0 locomotives in the classic Victorian style and it seems he developed a liking for that wheel arrangement. Upon moving south, he was faced with different traction problems and early designs for the L & SWR varied from the 0-4-4 'M7' to the 0-6-0 '700' Class goods engines. His preference did not remain dormant for long; in 1898 the 4-4-0 'C8' appeared, but it was not a success on main-line express trains, its 'M7' boiler not being up to the task. Drummond appears to have been well aware of the problem even before the 'C8' took to the rails, as he set in motion plans for the 'T9' with a much larger grate and firebox.

It was in their boiler dimensions that the 'C8' and 'T9' Classes mainly differed, and such was Drummond's confidence in the design that he ordered fifty locomotives straight from the drawing-board. Thirty were produced by Dubs & Co of Glasgow and twenty at the railway's own Nine Elms works. No sooner was that order complete than another 15 were ordered from Nine Elms and, in 1901, a final 'T9' was constructed by Dubs for the Glasgow exhibition of that year; this locomotive was subsequently sold to the L & SWR bringing the class total to 66.

Originally, the 'T9s' had 18.5 in diameter cylinders, and sand boxes in front of the leading coupled wheel splashers. The Dubs engines also had water tubes in the firebox but that feature did not extend to the first machines built at Nine Elms although it was incorporated in the final 15. Whether that feature was stipulated by Drummond or was a Dubs preference is unknown; it was functional, but complicated the firebox unnecessarily. The 'T9s' quickly established a reputation for speed and performance on the London & South Western's passenger expresses during their

Still in service after fifty years, 'T9' No 30301 at Salisbury on 24 November 1952 (D. K. Jones Collection).

early years in service, and the nickname 'Greyhound' was soon applied.

Robert Urie succeeded Drummond in 1912 and set about stamping his own authority on the railway, particularly in his modification of earlier classes and in the application of superheating. The 'T9's were performing well and authorization to modify them was not sought until the final years before the grouping of 1923. Urie had a few, new boilers constructed but mostly the original boilers were reconditioned. The firebox tubes were dispensed with, and new extended smokeboxes and stove-pipe chimneys, together with the removal of the sandboxes to a position between the frames, radically altered their appearance. Finally, the cylinder diameter was increased to 19 in.

As newer locomotives took over main-line express passenger duties, the 'T9s' proved to be excellent on secondary routes. Their relatively light axle loading and ability to negotiate reasonably tight curves proved invaluable. Not surprisingly, many of the class survived well beyond nationalization. Extension of the Southern Region's electrification scheme saw the last of the class withdrawn in 1960, but one, No 30120, was set aside for preservation. This locomotive, now part of the National Collection, has been restored to operational service on the Mid-Hants Railway, a glorious reminder of a long-lived and extremely useful class. Their work was not as glamorous as that of the main-line classes, but they had style and outlived most.

MIDLAND RAILWAY 'COMPOUND' 4-4-0

Designers Johnson/Deeley/Fowler. **Introduced** 1902/1905/1924. **Power classification** 4P. **Driving wheel diameter** 6 ft 9 in. **Leading wheel diameter** 3 ft 6 in. **Cylinders** One high pressure (inside) 19 in diameter × 26 in stroke, two low pressure (outside) 21 in diameter × 26 in stroke. **Valve gear** Stephenson. **Boiler**—*Pressure* 200 psi. *Grate area* 28.4 sq ft. **Heating surfaces**—*Tubes* 1170 sq ft. *Firebox* 147.3 sq ft. *Superheater* 290.7 sq ft. **Tractive effort** 22,650 lbs **Weight** Loco 61 tons 14 cwt, tender 42 tons 14 cwt. **Number built** 45 by Johnson and Deeley, 195 by Fowler. (The figures refer to the Fowler locomotives introduced in 1924.)

The Fowler 'Compounds' of 1924 were variations of an earlier arrangement introduced by Johnson on the Midland Railway in 1902 and subsequently modified by Deeley in 1905.

The use of 'compounding', steam successively expanded from a high pressure cylinder to a lower pressure in larger diameter cylinders, to improve efficiency had been employed in stationary and marine plant since the mid-1800s, but its application to railways presented problems. Space and complexity were the main restrictions, but a number of designers did attempt its use, mostly with little success.

On the Midland, S. W. Johnson developed a compound system based on a three-cylinder arrangement introduced in 1898 on the North Eastern Railway, and his 4-4-0 'Compound' was a major improvement which functioned effectively and efficiently. Independent reversing gears were provided for the high and low pressure cylinders but they could be operated from the same lever. It was also possible to admit high pressure steam to the low pressure cylinders if necessary for starting or hill climbing.

Deeley, CME of the Midland after Johnson, applied superheating and a number of other modifications to the basic compound design; trials had shown a reduction in coal consumption of more than 25% when superheating was employed. He subsequently rebuilt the Johnson machines in a similar manner, and with the typically light Midland express loads of the period the 'Compounds' proved themselves to be ideal.

Midland dominance of the LMS following the grouping made it inevitable that the 'Compounds' would be to the fore in the choice of express passenger engines. This proved to be the case, after trials were carried out on the Settle to Carlisle line in which the 'Compound' proved itself superior to the L & NWR 'Prince of Wales' Class. Even before the trials, an order for twenty superheated 'Compounds' had been placed—somebody at Derby was either confident or prejudiced! When the results of the trials proved 'Compound' superiority, large-scale production of the Fowler modified types could proceed.

Fowler modifications included a reduction in driving wheel diameter by 3 in to 6 ft 9 in and an increase in cylinder diameter of 0.75 in. Following the first batch of 1924, the cylinder dimensions reverted to the original and the enlarged units were subsequently modified. Production continued steadily until 1927 with the Vulcan Foundry and the North British Locomotive Co providing a number of locomotives. Fowler's successor, William Stanier, allowed the 1932 order for four locomotives to stand but production then ceased.

A Midland 'Compound' in BR days – No 41071 in 1955 (D. K. Jones Collection).

Preserved Midland 'Compound' No 1000 double-heads a railtour with 'Jubilee' class Leander.

Although the 'Compounds' were excellent locomotives for their size and type, they were quickly outpaced by the increase in traffic and train size. Double-heading became common with the resultant wasteful employment of men and machines. A 'Compound' was a more precise piece of machinery than a 'simple' and required careful driving; when taken away from top link duties to be operated by inexperienced drivers and firemen, their performance suffered. It is easy in retrospect to make critical remarks concerning the 'Compounds' but the main faults lay not with the machine but with the railway itself. 'Compounds', in effect, stayed too long on the express passenger scene, and outlived their time because of the narrow-minded attitude of those Midland men who dominated the LMS. In their day, however, they forcibly pointed out the benefits of compounding of which they were the ultimate British example. It remains a pity that the noble experiment was not developed. One example from those days, No 1000, remains, a rightful part of the National Collection.

GNR 'C1' CLASS 4-4-2

Designer H. A. Ivatt. **Introduced** 1902. **Driving wheel diameter** 6 ft 8 in. **Leading wheel diameter** 3 ft 8 in. **Trailing wheel diameter** 3 ft 8 in. **Cylinders** Two (outside) 20 in diameter × 24 in stroke. **Valve gear** Stephenson. **Boiler** (Figures given are for the original superheated boilers)—*Pressure* 170 psi. *Grate area* 31 sq ft. **Heating surfaces**— *Tubes* 2,360 sq ft. *Firebox* 141 sq ft. *Superheater* 430 sq ft. **Tractive effort** 17,340 lbs. **Weight** Loco 69 tons 12 cwt, tender 43 tons 2 cwt. **Number built** 94.

Ivatt introduced the 'Atlantic' type to Britain in 1898, narrowly beating Aspinall whose 4-4-2 entered service with the Lancashire & Yorkshire Railway the following year. There was a need on the Great Northern for higher-powered locomotives to take over the duties of the ageing Stirling 'singles' but it appears likely that No 990, Ivatt's first 4-4-2, was hurriedly introduced in order to beat Aspinall. It was some years before quantity production took place at Doncaster, but the L & Y had twenty of its 'Atlantics' in service before the end of 1899. Despite that, the Ivatt machine marked a radical change for the GNR and heralded a new era of high-powered locomotives. The nominal tractive effort of No 990 was actually less than the final batch of Stirling 8 ft 'singles' but the large-capacity boiler allowed for a better, more consistent performance.

Ivatt was not content to leave matters there and soon busied himself with improvements. In 1902, his 'large Atlantic' No 251 appeared. It was built with the same chassis, cylinders and valve arrangements as No 990 but, aware of the need to provide sufficient steam, Ivatt provided it with a boiler almost to the limit allowed by the loading gauge. The total heating surface was some 900 sq ft greater than that of the original 'Atlantic' design and an extra 6.5 sq ft of grate area would provide the heat. The boiler gave the locomotive an attractive and powerful appearance but that was not matched by its performance; in effect, No 251 and her sisters were overboilered for the small 19 in diameter cylinders. Had Ivatt increased the cylinder size in line with the boiler capacity, he would have produced a remarkable engine for the period rather than just a very good one.

Superheaters were provided on the final ten members of the class constructed in 1910, and 20 in diameter cylinders with piston valves were fitted; the earlier locomotives were subsequently rebuilt to that form. This improved their performance but the

Above *Three 'Classic Locomotives' at the National Railway Museum, York. Left to right: Southern Railway 'Schools' Class* Cheltenham, *Midland 'Compound' No 1000 and Stirling 'Single' No 1. On the extreme right is LB & SCR* Gladstone.

Below *Preserved GCR 'Director'* Butler Henderson *departing from Loughborough Central.*

Above *The Festiniog Railway's newer 'Double Fairlie'* Earl of Merioneth *at Porthmadog.*

Below *Ex-LB & SCR 'Terrier' No 32670 on loan to the Great Western Society, Didcot, from the Kent & East Sussex Railway.*

Above *'Jinty' No 47383 with a demonstration freight train at Highley on the Severn Valley Railway.*

Below *'The Flying Pig', Ivatt Class '4' No 43106, and WD 'Austerity' 2-10-0 Gordon at Bridgnorth in June 1984.*

Above *Stanier '8F' No 48423 at Rose Grove sheds on 13 April 1968* (Chris Surridge).

Below *'Black Fives' Nos 45290 and 45110 at Bolton sheds on 12 April 1968* (Chris Surridge).

Ivatt 'C1' Class large-boilered 'Atlantic' No 4418 at Doncaster on 17 September 1926 (National Railway Museum, York).

nominal tractive effort remained the same as previously because the boiler pressure was reduced to 150 psi, as Ivatt considered that such action would reduce boiler maintenance. Gresley took over from Ivatt the following year and one of his early tasks was to fit superheaters to the remaining large-boilered 'Atlantics', and, if cylinder replacement was necessary, new 20 in diameter units were provided. Instead of reducing the boiler pressure, Gresley kept it at 170 psi which became standard throughout the class. The improvement in performance and economy was immediately evident.

Loads of 500 or even 600 tons were taken on the steep gradients out of King's Cross and the 'C1' 'Atlantics' justly earned their reputation. Gresley continued their development by fitting even larger superheaters and that allowed the engines to be worked harder without any deterioration in performance or steam production. Although their capabilities were favoured by the crews, cab facilities and smoothness of ride left much to be desired. There was barely enough space for the fireman to swing his shovel and riding was notoriously rough. This was attributed to the lack of side control on the trailing wheels which were provided with a large amount of lateral play, but despite this they were popular locomotives.

In these machines, Ivatt provided the groundwork upon which Gresley could build his remarkable large-boilered locomotives during the years which followed. They were classics because of what they achieved and what they led to. The original large-boilered 'Atlantic', No 251, now restored to its Ivatt external condition, is displayed at the National Railway Museum in York.

GWR 'CITY' CLASS 4-4-0

Designers William Dean/G. J. Churchward. **Introduced** 1902. **Driving wheel diameter** 6 ft 8.5 in. **Leading wheel diameter** 3 ft 8 in. **Cylinders** Two (inside) 18 in diameter × 26 in stroke. **Valve gear** Stephenson. **Boiler** (Figures given are for the superheated boiler as subsequently fitted)—*Pressure* 200 psi. *Grate area* 20.56 sq ft. **Heating surfaces**— *Tubes* 1,350 sq ft. *Firebox* 129 sq ft. *Superheater* 192 sq ft. **Tractive effort** 17,790 lbs. **Weight** Loco 55 tons 6 cwt, tender 40 tons. **Number built** 20.

Churchward is generally given credit for the design of the 'Cities' but that is not strictly correct. It is true that the main batch appeared in 1903 after he had assumed control of the Locomotive Department, but the class itself was based upon a 4-4-0 design of 1900. In September 1902, Churchward simply took 'Atbara' Class locomotive *Mauritius* and fitted it with his new Standard No 4 boiler. Thus the 'Cities' were born, below the frame pure nineteenth-century Dean and above pure twentieth-century Churchward. It was a combination which proved to be highly successful and certainly filled an express passenger gap whilst Churchward developed his standard range of locomotives.

The year 1903 saw the construction of ten identical locomotives at Swindon. These were the real 'Cities', being new engines and actually named after cities in England. Amongst that batch was *City of Truro* whose exploits on Wellington Bank are now part of locomotive history. On May 9 1904, whilst hauling a boat train from Plymouth to Paddington, a speed of 102.4 mph was achieved on the descent of that incline. At least, that was the speed according to the timings of the eminent observer Charles Rous-Marten. Over the years, the exploit has been questioned and many have contended that *City of Truro* was not, therefore, the first British locomotive to exceed 100 mph. Without evidence to firmly contradict the timings, it is impossible to make a definite decision on the matter, but what is certain is that it was a very fast run indeed and the 'Cities' became famous for their high-speeds.

A further nine members of the class were constructed from 'Atbaras' between 1907 and 1909, and all the 'Cities' were provided with superheaters by 1912. During their early years of service, the 'Cities' almost monopolized the main expresses between Paddington and Cornwall, but, as more of Churchward's 'Saints' and 'Stars' appeared, they were gradually replaced. Service on expresses to the Midlands and then South Wales

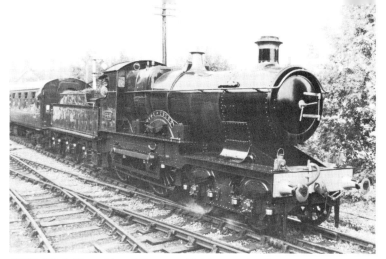

GWR 'City' Class City of Truro *at Highley on the Severn Valley Railway during 1986.*

followed, but the growing ranks of Churchward standard locomotives saw them displaced from each of these services. Being, to a great extent, of older design, their fate was always to be replaced by newer machines.

This in no way detracted from their importance or abilities—it is just that they came at the wrong time. It was the rapid locomotive developments initiated by Churchward which overtook them, rather than any shortcomings of their design. Had they been allocated to other railways they would almost certainly have lasted longer on major express trains. Even so, they were not finally withdrawn from service until 1930 which, considering the major locomotive revolution on the Great Western during that period, is a very good achievement for what was essentially a nineteenth-century design.

The exploits attributed to *City of Truro* make it apropriate that this locomotive should be preserved, but the class as a whole deserves to have been better represented in preservation. The 'Cities' marked an evolutionary change from nineteenth to twentieth century locomotive design and they are an important link with the past. The remarkable thing is that *City of Truro* owes its preservation to the LNER, not the Great Western. The authorities at Paddington were quite prepared to scrap the famous engine, not considering it to be of outstanding importance. Fortunately, the LNER stepped in with the offer of a place in its museum at York. Since that date in 1931, the locomotive has seen two periods of service divided by a spell at the Great Western Museum, Swindon.

GWR 'SAINT' CLASS 4-6-0

Designer G. J. Churchward. **Introduced** 1903. **Power classification** 4P. **Driving wheel diameter** 6 ft 8.5 in. **Leading wheel diameter** 3 ft 2 in. **Cylinders** Two (outside) 18.5 in diameter × 30 in stroke. **Valve gear** Stephenson. **Boiler**—*Pressure* 225 psi. *Grate area* 27.1 sq ft. **Heating surfaces**—*Tubes* 1,687 sq ft. *Firebox* 154 sq ft. *Superheater* 283 sq ft. **Tractive effort** 24,395 lbs. **Weight** Loco 72 tons, tender 40 tons. **Number built** 77. (The figures given are for the final batch of 1913 to which form many of the earlier locomotives were converted.)

Churchward's standardization scheme called for a two-cylinder 4-6-0 locomotive, and No 98 appeared in 1903. Another 4-6-0, No 100, had been produced at Swindon a year earlier but there is doubt as to the extent of the influence of William Dean on the design, Dean having been CME until May 1902. No 98 can, however, be described as pure Churchward. It was not a final design, only a prototype, but there were features which found use in all two-cylinder designs of the Churchward era. Most notable was the cylinder block which employed two identical castings bolted back-to-back to form the cylinder and valve casing together

Churchward 'Saint' Class No 181 Ivanhoe *in its original form as an 'Atlantic'* (Flewellyn Collection, GWR Museum, Swindon).

with the smokebox saddle. Churchward's preference for horizontal cylinders and a dislike of visible valve gear also showed in the design.

A second locomotive of the type, No 171, left Swindon works later that year, but this had a boiler pressure of 225 psi instead of the earlier 200 psi. Following his trials with the French-built 'Atlantics', Churchward decided to convert No 171, later known as *Albion*, to that form. Further 'Saints', as the class became known, were constructed during 1905, some being 4-6-0s and others 4-4-2s. A year of testing proved that the 4-6-0 wheel arrangement was the most suitable and further construction was to that form, with all the 'Atlantic'-type 'Saints' being subsequently rebuilt.

The 'Saints' also provided a mobile test bed for superheating, with different types being tried on various members of the class. Results showed that superheating was advantageous and also proved, at least to Churchward, that his own development, the Swindon No 3 superheater, was more suitable than any other form. All members of the class had been fitted with superheated boilers by 1912 and the final batch of 'Saints' left Swindon works during the early months of 1913.

In express passenger terms, the class was not large, only 77 strong, but the 'Saints' were highly significant due to the part they played in GWR locomotive development. It was their introduction which showed the way to a modern passenger locomotive fleet, whilst comparisons between wheel arrangements illustrated the superiority of the original 4-6-0 formation. The superheater trials could only have been effectively carried out with a class which was otherwise performing uniformly to high standards, and the 'Saints' were certainly doing that on expresses to Devon and Wolverhampton. So useful was the class that Collett used the design as a basis for his highly versatile and successful 'Hall' class when, in 1924, *Saint Martin* was fitted with smaller diameter wheels and proved to be a perfect mixed traffic engine. Ultimately, 258 'Halls' were constructed to the basic 'Saint' design and a further 71 to a modified design worked out by Hawksworth. The fact that production of the fundamental 'Saint' design continued until 1943 illustrates how useful Churchward's two-cylinder 4-6-0 pattern was.

Unfortunately, no 'Saint' has survived, but a large number of 'Halls' have and there are plans, finance permitting, to convert one of that number into a 'Saint'—rebuilding in reverse!

GWR '2800' CLASS 2-8-0

Designer G. J. Churchward. **Introduced** 1903. **Power classification** 8F. **Driving wheel diameter** 4 ft 7.5 in. **Leading wheel diameter** 3 ft 2 in. **Cylinders** Two (outside) 18.5 in diameter × 30 in stroke. **Valve gear** Stephenson. **Boiler** (Figures given are for the standard superheated type)—*Pressure* 225 psi. *Grate area* 27.22 sq ft. **Heating surfaces**— *Tubes* 1,608 sq ft. *Firebox* 150 sq ft. *Superheater* 265 sq ft. **Tractive effort** 35,380 lbs. **Weight** Loco 75 tons 10 cwt, tender 43 tons 3 cwt. **Number built** 166.

When Churchward produced his scheme for six standard locomotive classes in 1901, it included one for heavy freight haulage. Introduction of prototype No 97 two years later heralded not only a new class but also a new wheel arrangement on the railways of Britain, that locomotive being the first 2-8-0 to operate in this country. In keeping with the standardization scheme, the cylinders and boiler were the same as those used for the prototype 4-6-0. Churchward could not be faulted in terms of the design or the use of standard parts, for his 2-8-0 was ideally suited to the traffic demands and its manufacture was easier using a limited number of parts standard across many classes.

With the success of No 97 readily apparent after test running, large-scale production could begin. The first production batches appeared in 1905 with only minor changes from the prototype. The piston valve diameter was increased from 8.5 in to 10 in and boiler pressure was raised by 25 lbs to 225 psi. The tender was also given an increased water capacity of 4,000 gallons. With the 1907 batch, it was decided to increase the cylinder diameter from the original 18 in, but not until the 1909 batch did it reach the final diameter of 18.5 in. Development of the Swindon No 3 superheater had followed trials with other classes, and the benefits produced dictated that superheating should be applied to the '2800' Class; this resulted in a tractive effort some 550 lbs greater than that of the prototype.

In service, the class showed itself to be capable of hauling the heaviest loads without undue difficulty, but towards the end of Churchward's term in office it became apparent that an enlarged version would be useful. Such a locomotive would have a greater reserve of power and be capable of a higher sustained speed. Churchward's final essay in locomotive design was, in fact, such a machine, and the new '4700' Class of 1919 could be considered as an enlarged version of the '2800s' At the same time, further

'2800s' were being constructed, but these had outside steam-pipes because sealing problems had arisen with the inside arrangement. Most of the earlier locomotives were later given outside steam-pipes when replacement cylinder castings were fitted.

Collett recommenced production of the class in 1938 after a period of 19 years. The new locomotives were practically identical to those produced previously (there were minor differences and a slight increase in weight) and there can be no better proof as to the soundness of the original design than that construction should start again 35 years after introduction. Obviously influenced by the freight transport needs of war-torn Britain, production continued until 1942. Apart from the limited number of '4700' Class locomotives, the GWR had no need for another freight design and neither of Churchward's successors had any desire or need to make other than basic modifications. As part of the 'coal for export' campaign after the war, some members of the class were fitted with oil-burning equipment but that was removed within two years because of an oil shortage.

Most of the class lasted until the 1960s, giving them a working life of 60 years—only good designs last that long. Rescued from Woodham's scrapyard at Barry, a number of '2800s' now operate on the preserved lines of Britain and few would argue against their reaching a century in service.

'2800' Class locomotive No 3822 as preserved at Didcot.

GWR 'STAR' CLASS 4-6-0

Designer G. J. Churchward. **Introduced** 1906. **Power classification** 5P. **Driving wheel diameter** 6 ft 8.5 in. **Leading wheel diameter** 3 ft 2 in. **Cylinders** Four, 15 in diameter × 26 in stroke. **Valve gear** Walschaerts inside with rocking levers to outside cylinders. **Boiler** (Figures given are for the standard superheated boiler)—*Pressure* 225 psi. *Grate area* 27.1 sq. ft. **Heating surfaces**—*Tubes* 1,599 sq. ft. *Firebox* 155 sq. ft. *Superheater* 260 sq ft. **Tractive effort** 27,800 lbs. **Weight** Loco 75 tons 12 cwt, tender 40 tons. **Number built** 73.

Churchward's standardization scheme of 1901 included a 4-6-0, but this was a two-cylinder design which subsequently became the 'Saint' Class. Trials with the French-built four-cylinder compound 'Atlantics' soon set him thinking along the lines of a four-cylinder simple. In 1906, using a number of features from the French engines, he produced No 40 which has since been considered as a turning point in British steam locomotive design.

This four-cylinder simple was actually constructed as a 4-4-2 in order to allow direct comparisons with the imported 'Atlantics' and his own two-cylinder versions. Within months, the matter had been settled with the 4-6-0 wheel arrangement being preferred and a four-cylinder simple being the ideal for high-power performance, and thus the 'Stars' were born.

The first production 'Stars' entered service during 1907, and No 40 was later rebuilt as a 4-6-0 and renumbered 4000. They were actually given the names held by the former broad gauge 'Star' Class engines produced by Robert Stephenson & Co, and the use of those names was probably something of a sop to the many enthusiasts who still blamed Churchward for authorizing the scrapping of the two preserved broad gauge engines.

The production 'Stars' differed from No 40 *North Star* in that they were provided with inside Walschaerts valve gear rather than the 'scissors' arrangement used on the prototype. The compensated rocking levers for operating the outside cylinder valves were, however, retained. Churchward disliked the idea of outside valve gear and inclined cylinders with the result that none of his designs made use of such features even when they might have made for improved accessibility and easier maintenance. The long taper boiler was a masterpiece and became the Standard No 1 as used for other classes.

Further batches of 'Stars' appeared almost annually until 1914 with minor improvements being made as they became necessary.

Churchward 'Star' Class No 4067 Tintern Abbey (GWR Museum, Swindon).

Superheating became an important aspect of the Churchward locomotive policy and, following extensive trials, he settled upon a relatively low superheat temperature compared with other railways, using his own standard Swindon No 3 type. All members of the class built after 1910 were constructed with superheaters and earlier engines ultimately converted. One of the 1913 batch, No 4041 *Prince of Wales*, was provided with 15 in diameter cylinders, thus raising the tractive effort. Doubts were expressed about the boiler's ability to successfully steam four 15 in diameter cylinders, but they proved groundless and the entire class was eventually fitted with the larger units.

At the time, the 'Stars' were acknowledged as the best express passenger locomotives on the Great Western, and possibly in Britain. Certainly the running of *Polar Star* on L & NWR metals during 1910 left no doubt as to the superiority of Churchward's machine over the best of that railway's stock. Further 'Stars' were constructed during 1922 and 1923 but after that Collett extended the design concept with his 'Castle' and later 'King' Classes. A number of 'Stars' were rebuilt as 'Castles' in later years.

In the 'Stars', Churchward had not only produced an excellent locomotive for the GWR but he had provided inspiration for other designers to follow. It is not easy to isolate one locomotive design from the complete Churchward standardized fleet, but the 'Stars' certainly stand out from the others, and the ancestry of many subsequent express passenger designs can be traced back to their introduction.

L & YR 'DREADNOUGHT' CLASS 4-6-0

Designer George Hughes. **Introduced** 1908. **Power classification** 5P. **Driving wheel diameter** 6 ft 3 in. **Leading wheel diameter** 3 ft 3 in. **Cylinders** Four, 16.5 in diameter × 26 in stroke. **Valve gear** Walschaerts. **Boiler**—*Pressure* 180 psi. *Grate area* 29.6 sq ft. **Heating surfaces**— *Tubes* 1,729 sq ft. *Firebox* 180 sq ft. *Superheater* 504 sq ft. **Tractive effort** 28,800 lbs. **Weight** Loco 77 tons 9 cwt, tender 40 tons. **Number built** 75 (includes original non-superheat types). (Figures given are for the final form produced in 1924-25.)

The Lancashire and Yorkshire was one of the few major railways to operate completely away from the London area, but it was nonetheless very progressive. Cross-Pennine passenger services were of a stop-start nature on heavily-graded track with the result that powerful and fast locomotives were required. In 1908, Hughes produced a four-cylinder 4-6-0 locomotive design and twenty were quickly ordered from the L&YR works at Horwich. These machines were massive by contemporary standards and soon earned the nickname 'Dreadnoughts' after the German battleships then being produced. Axle loading was high at 19.8 tons and they were barred from NER track east of Goole. Performance also left a lot to be desired as it did not live up to design expectations. They had been intended to provide rapid acceleration with heavy trains but failed to do so, and although Hughes was nothing if not inventive, little could be done to improve the stock immediately.

Hughes had enthusiasm for superheating but favoured a higher degree than that advocated by Churchward on the GWR. Following trials with superheaters on some 4-4-0s, he asked the Horwich drawing-office to prepare plans for a superheated boiler to fit the 'Dreadnoughts', but it was not until the end of the First World War that any construction could take place. By 1919, most of the class was laid up awaiting repairs and the opportunity presented itself for conversion. No 1522 was first in line, receiving the new superheated boiler with a modified ashpan as well as a redesigned front end. The improvement was almost beyond belief, resulting in a decision to not only rebuild the entire class but also construct further 'Dreadnoughts' to the new design. These additional 55 locomotives appeared between 1921 and 1925, the latter batch entering LMS service following the grouping of 1923. It was, in fact, a member of the class, LMS No 10433, which actually

Above *Former L & YR 'Dreadnought' 4-6-0 as LMS No 10452 in April 1927* (R. D. Stephenson Collection, NRM, York).

Below *Former L & YR 'Dreadnought' at Blackpool on 2 September 1938, LMS No 10442* (D. K. Jones Collection).

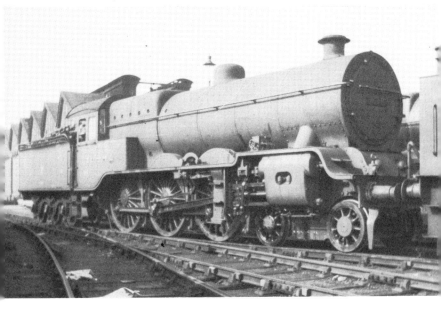

became the first LMS locomotive, being released for traffic on January 1 1923.

Following the grouping, many members of the class were allocated to sheds in the northern districts and used for hauling express trains over Shap. Their power and hauling capacity could not be faulted but they were very heavy on coal, and compared with the former L & NWR 'Claughtons' or Midland 'Compounds' they were grossly uneconomical, although their power remained in demand. Major withdrawals began as the Stanier passenger locomotives entered service during the early 1930s but it was not until 1951 that the final member of the class was withdrawn. Regrettably, nobody seems to have considered preserving the remaining example of this large Lancashire & Yorkshire passenger design.

It is unfair to judge the Hughes 'Dreadnoughts' by comparison with equivalent L & NWR or Midland locomotives, because they were designed for different purposes. L&YR passenger operations called for locomotives capable of fast running between stations only a short distance apart, and that is what the Hughes 4-6-0s were intended to achieve. They were not produced for long non-stop runs over the Lakeland hills, although comparison trials were carried out in that terrain and on that type of running. In their superheated modified form, the 'Dreadnoughts' were ideal machines for the Lancashire and Yorkshire express passenger services. That was what they were designed for, a definite case of 'horses for courses'.

L & NWR 'PRINCE OF WALES' CLASS 4-6-0

Designer C. J. Bowen-Cooke. **Introduced** 1911. **Power classification** 4P. **Driving wheel diameter** 6 ft 3 in. **Leading wheel diameter** 3 ft 9 in. **Cylinders** Two (inside) 20.5 in diameter × 26 in stroke. **Valve gear** Joy. **Boiler**—*Pressure* 175 psi. *Grate area* 25 sq ft. **Heating surfaces**—*Tubes* 1,376 sq ft. *Firebox* 136 sq ft. *Superheater* 304 sq ft. **Tractive effort** 21,700 lbs. **Weight** Loco 66 tons 5 cwt, tender 39 tons 5 cwt. **Number built** 245.

George Whale's 4-6-0 'Experiment' Class had performed with moderate success on the L & NWR's lines from Euston since its introduction in 1905. By the end of that decade, superheating had become the accepted path towards more economical performance and Bowen-Cooke was a committed believer. However, the Civil Engineer would not accept the weight diagrams of the authorized four-cylinder 4-6-0 design of 1911 and the works order was quickly changed to a batch of superheated 'Experiments'. Changes in design were made, particularly with cylinder size and the use of 8 in diameter piston instead of slide valves, and that batch became the first of many 'Prince of Wales' Class locomotives.

The class name followed from the fact that the investiture of the Prince of Wales, later King Edward VIII, was being held at Caernarvon in 1911. The ten class members were constructed in that year and, although designed as express passenger locomotives, they soon showed an ability as mixed traffic engines. Their maximum speed was below that of the 'Experiments' but their economy was higher. Despite this, however, there was no desire to superheat the 'Experiments' in order to put them on a par with the 'Princes'.

A further 16 locomotives were built at Crewe in 1913 with 14 more entering service the following year. Production continued at Crewe but the First World War placed a considerable strain on the works and an order for twenty of the class went to the North British Locomotive Co in 1915. After the war, William Beardsmore & Co received an order for no fewer than ninety locomotives and sixty tenders; the contract, worth £90,000, was the largest to be placed with a private builder by any railway up to 1923.

Prior to the grouping, these 4-6-0s were the preferred form of power for hauling passenger trains on the heavy gradients

L & NWR 'Prince of Wales' Class, LMS No 5040.

between Lancaster and Carlisle, but early in LMS days trials were carried out to compare locomotives from the different constituent companies. Although they did well in terms of haulage and time-keeping, the 'Princes' could not match the Midland 'Compounds' in terms of coal consumption. Midland dominance of the LMS resulted in, amongst other things, the 'Compounds' being chosen for express duties while the 'Princes' were relegated. Troubles with pin wear on the indirect drive Joy valve gear had been experienced; it was modified to direct action but it failed to cure the problem. The use of Belpaire boilers was tried but it was more a means to standardizing boilers than an attempt to modernize the class. Several modernization schemes were proposed involving new cylinders and valve gear; there was even a proposal to fit Caprotti valve gear. That nothing happened is probably a symptom of the Midland's domination of the LMS. With the arrival of Stanier the end was in sight, and construction of his 'Black 5s' allowed rapid withdrawal; a few survived the Second World War before finally succumbing in 1949.

The 'Princes' were equal to their designed duty as express passenger locomotives and only changing circumstances resulted in their potential remaining unfulfilled. As a test bed for superheating, they provided invaluable experience and, although unspectacular, they were essential to locomotive development.

MIDLAND RAILWAY CLASS '4' 0-6-0

Designer Henry Fowler. **Introduced** 1911. **Power classification** 4F. **Driving wheel diameter** 5 ft 3 in. **Cylinders** Two (inside) 20 in diameter × 26 in stroke. **Valve gear** Stephenson. **Boiler**—*Pressure* 175 psi. *Grate area* 21 sq ft. **Heating surfaces**—*Tubes* 1,034 sq ft. *Firebox* 123 sq ft. *Superheater* 253 sq ft to 313 sq ft. **Tractive effort** 24,560 lbs. **Weight** Loco 48 tons 15 cwt, tender 41 tons 4 cwt. **Number built** 772.

Mineral traffic was a major part of the Midland Railway's business whilst Midland routes were subject to severe weight restrictions and width limitations, so it was not easy to avoid double-heading with small locomotives. By 1910, it was becoming obvious that a replacement for the fairly outdated 0-6-0 goods engines was required, but that wheel arrangement was still the only possibility considering the route restrictions. Superheating was in vogue at the time and Fowler decided to try out a modernized version of the standard 0-6-0 but with superheating. Two locomotives appeared in 1911, each fitted with a different design of superheater.

A period of extensive trials ensued, and it was not until 1917 that the next batch of 15 engines appeared, though possibly the war had something to do with the delay. Further locomotives were produced at Derby and by Armstrong-Whitworth in the years up to 1922, five of which were built especially for work on the Somerset & Dorset Railway. Operational reports from the two prototypes and further large-scale construction undertaken between 1924 and 1928 confirmed the belief that the design was a good one. Over 500 locomotives were produced at the major works and by outside contractors like the North British Locomotive Co, Andrew Barclay and Kerr Stewart, and the design was designated as a standard for the newly-constituted LMS.

Fowler made several minor modifications in order that the design might suit the larger LMS system although some defects such as the undersized axle-box bearings were perpetuated. For LMS operations, vacuum brakes became standard as did steam heating hoses for use with passenger trains. Several different sizes of boiler were used though the variation in dimensions was not major except with regard to the superheater area which varied widely from batch to batch. Further batches were authorized by Stanier between 1937 and 1941. He himself had proposed a modernized 0-6-0 and even a 2-6-0, but the Operating

Above *Fowler '4F' No 4027 at the Midland Railway Centre.*

Below *'4F' 0-6-0 No 43924 at Haworth on the Worth Valley Railway.*

Department indicated a satisfaction with the 0-6-0 '4F' which suited its purposes admirably.

Freight locomotives have always been looked upon less favourably compared with the larger, more glamorous passenger engines but their work was just as useful, perhaps more so during times of conflict. Working much of their time away from main lines, these little locomotives performed valuable service away from the public gaze. The fact that the railway's Operating Department expressed a preference for this relatively old design rather than opting for a new Stanier locomotive must indicate more than mere satisfaction; at the time, 1937, Stanier's reputation was at its highest.

The '4Fs' were not as efficient as later freight locomotives but they were of a simple and robust design which could withstand the neglect of post-war years without resulting in drastic failures. This hard-working, go-anywhere (almost) class lasted in British Railways, service until 1966, 55 years after its first introduction, and outlived many later designs, an achievement of which only a first-rate design would have been capable. Provision of vacuum braking and steam heating turned them into useful passenger engines with a good turn of speed, and whilst it would not be accurate to describe them as 'mixed traffic', they were certainly maids of all work.

A number have been preserved, allowing the present generation to witness their capabilities on passenger trains.

GCR 'DIRECTOR' CLASS 4-4-0

Designer J. G. Robinson. **Introduced** 1913. **Power classification** 3P2F. **Driving wheel diameter** 6 ft 9 in. **Leading wheel diameter** 3 ft 6 in. **Cylinders** Two (inside) 20 in diameter × 26 in stroke. **Valve gear** Stephenson. **Boiler**—*Pressure* 180 psi. *Grate area* 26.5 sq ft. **Heating surfaces**—*Tubes* 1,502 sq ft (1,388 sq ft). *Firebox* 157 sq ft (155 sq ft). *Superheater* 303 sq ft (209 sq ft). **Tractive effort** 19,644 lbs. **Weight** Loco 61 tons 3 cwt, tender 48 tons 6 cwt. **Number built** 10 ('D10' Class) 35 ('D11 Class). (Figures given in brackets are for the 'D11' Class 'Improved Directors' of 1919.)

The 'Sir Sam Fay' Class 4-6-0 locomotives had been introduced for Great Central main-line duties in 1912 and it seemed strange that a 4-4-0 of marginally lower tractive effort should be constructed less than a year later. The Great Central '11E' Class, later 'D10' under LNER classification, became more commonly known as the 'Director' Class as the ten locomotives were named after GCR board members. The boiler was shorter than that of the 4-6-0s but had a similar grate area and, because the wheel arrangement allowed an unrestricted air supply, it turned out to be a much better steamer. These 'Directors' eventually took over the hardest GCR passenger turns which included the through workings between London and Manchester.

The locomotive was attractive in style and looked impressive because of the bold spacing between the coupled wheel axles. Outside admission piston valves were employed, these being actuated by Stephenson gear through an offset arm. That was an unconventional way of operating the valve spindle but was chosen because it would allow for easy conversion to inside admission without changing the eccentrics. With subsequent 'Directors', inside admission valves were fitted and the earlier engines were converted merely by substituting a rocking lever valve spindle drive in place of the offset arm.

In 1919, a further eleven 'Directors' were constructed, but these differed from the earlier examples in a number of ways. One of these related to the cab, but the boiler was the main difference between the groups. On the 'Improved Directors' it had a reduced heating surface area but the grate area remained the same, thereby allowing for better steam generation qualities. The boiler was also set higher in the frames and so required a shorter chimney and dome in order to meet loading gauge restrictions. If anything, it was considered that the 'Improved Directors' were

The only surviving 'Director'; Butler Henderson on the Great Central Railway at Loughborough, 7 September 1986.

more visually impressive than their earlier sisters.

In terms of power they were also more impressive and became the mainstay of the Great Central's express passenger fleet. It was not until the mid-1930s that Gresley locomotives began to displace them from the principal services between Marylebone and Manchester. Very high speed was not a necessity for GCR services, but the 'D11s' could manage 70 mph with ease and at times reached 90 mph. In 1924, a further 22 'D11s' were constructed by the LNER for service in Scotland; they never operated on former Great Central metals. They were identical in all respects save for dimensional changes to suit the Scottish loading gauge. Whilst the Great Central 'Improved Directors' were blessed with impressive names like *Ypres*, *Somme* and *Jutland*, the Scottish engines carried plates announcing *Wizard of the Moor*, *Laird of Balmawhapple* and *Luckie Mucklebackit*.

As representatives of the Edwardian era, these engines were impressive both in terms of style and power. It is appropriate, therefore, that the first of the 'D11s' has been restored to working order and now operates on its old ground at the Great Central Railway, Loughborough.

S & DJR CLASS '7F' 2-8-0

Designer Henry Fowler. **Introduced** 1914. **Power classification** 7F. **Driving wheel diameter** 4 ft 7.5 in. **Leading wheel diameter** 3 ft 3.5 in. **Cylinders** Two (outside) 21 in diameter × 28 in stroke. **Valve gear** Walschaerts. **Boiler** (Figures in brackets are for the second batch of five built 1925)—*Pressure* 190 psi. *Grate area* 28.4 sq ft. **Heating surfaces**—*Tubes* 1,180 sq ft (1,323 sq ft). *Firebox* 147.25 sq ft (148 sq ft). *Superheater* 290.75 sq ft (374 sq ft). **Tractive effort** 35,392 lbs. **Weight** Loco 64 tons 15 cwt, tender 44 tons 4 cwt. **Number built** 11.

The Midland Railway had responsibility for locomotives and rolling-stock on the Somerset & Dorset Joint Railway under the 1876 lease agreement with the London & South Western Railway. The line was unlike anything else under Midland control in that there were long gradients of up to 1 in 50, and the heavy mineral traffic resulted in short trains and the necessity of banking. By 1911, the situation had become serious and the Locomotive Superintendent of the line once more petitioned Derby for more power; this time his request met with a response.

Fowler set his drawing office to work producing a 2-8-0 based upon an earlier proposed 0-8-0. The use of outside cylinders with Walschaerts valve gear was a fundamental departure from normal Midland practice with regard to mineral engines. Although Fowler is given credit for the design, it should really go to the drawing office team under the control of James Clayton, and the resulting locomotive looked impressive and, to some people, even attractive. In its original form the tender was provided with a cab which matched that of the locomotive; this was not only smart but functional, as a good deal of tender-first running could be expected due to the lack of turntables on the S & DJR. Token exchange apparatus was also provided on both sides because of the extensive mileage of single-line track on the railway. Heavy freight working and the nature of the route necessitated superior braking ability; three brake cylinders were fitted and all wheels had brakes, including those on the pony truck. It was definitely a design for working a specific route.

In operation, the class proved capable of handling any traffic the line could provide and in 1925 a further five engines were constructed to add to the original six (for such a short line only a small class was required). The second batch, constructed by Robert Stephenson & Co rather than at Derby, had a number of modifications, especially to the boiler. This was larger than the original and non-standard to the Midland, but ultimately all the locomotives were fitted with the original type of boiler. Stanier made several changes including the removal of the pony truck brake and an increase in the coupled wheel diameter of 1 in by fitting new tyres.

The class was also used for passenger duties, this being a regular feature of their operation from 1950 until the line was closed. No other British freight locomotive was employed so extensively as a policy on passenger trains, and the S & D 7Fs were equally at home on such duties, even double-heading as the need arose. As a design, they were the ideal 'horse' for the Somerset & Dorset's 'course', and impending closure brought thousands of enthusiasts to the line, as much to see the unique 2-8-0s in action as to see the route itself.

Fortunately, two of the class have been preserved, one, No 13809, approved for main-line running where its outstanding performances still bring out the crowds. The uniqueness of these machines would be enough to make them classics, but their record in service confirms their place in locomotive history—for a period of fifty years they more than fulfilled their design requirements.

S & DJR '7F' No 53806 at Swindon on 10 March 1963 (D. K. Jones Collection).

HIGHLAND RAILWAY 'CLAN' CLASS 4-6-0

Designer Christopher Cumming. **Introduced** 1919. **Power classification** 4P. **Driving wheel diameter** 6 ft 0 in. **Leading wheel diameter** 3 ft 3 in. **Cylinders** Two (outside) 21 in diameter × 26 in stroke. **Valve gear** Walschaerts. **Boiler**—*Pressure* 175 psi. *Grate area* 25.5 sq ft. **Heating surfaces**—*Tubes* 1,328 sq ft. *Firebox* 139 sq ft. *Superheater* 256 sq ft. **Tractive effort** 23,688 lbs. **Weight** Loco 62 tons 5 cwt, tender 42 tons 1 cwt. **Number built** 8.

Few people realize that Scotland once had an extensive railway system through some of the most difficult terrain in Britain. The Highland Railway commenced at Stanley Junction just north of Perth and extended to Thurso at the northern tip of the mainland and Kyle of Lochalsh to the west. Although the traffic may not have matched some of the English lines, it was necessary to provide large and powerful locomotives for these long runs over lines with severe gradients.

The 'Castle' Class of 1900 was satisfactory for express passenger duties but in 1914 Frederick Smith, the Chief Mechanical Engineer, produced a design for a more powerful engine to cope with the increasingly heavy traffic between Perth and Inverness. By chance, an error in a draughtsman's drawing was missed and the locomotives produced by Hawthorn, Leslie & Co turned out to be too heavy. The Civil Engineer refused to accept them and Smith was called upon to resign, which he did. These 'River' Class engines were then sold to the Caledonian Railway, some people say at a profit, but the fact was that the Highland remained short of powerful engines.

Christopher Cumming replaced Smith and proceeded to build up the stock with construction of further 'Castles' together with some 4-4-0 and 4-6-0 goods engines of his own design. Wartime restrictions prevented construction of passenger locomotives unless specifically sanctioned by the Railway Executive Committee, so in 1917 the Highland approached that body for permission to construct four superheated 4-6-0 passenger engines. The route north was extremely important due to the location of the Grand Fleet at Scapa Flow and these locomotives were urgently required. Permission was granted and a design was worked out with Hawthorn, Leslie & Co for an engine based on the Cumming 4-6-0 goods locomotive but with large wheels and boiler. Because of the

Former Highland Railway 'Clan' in LMS days; No 14764 Clan Munro *at Aviemore in about 1947* (F. Carrier Collection, NRM, York).

fiasco with the 'River' Class, the Civil Engineer was also called upon to approve the design. It is difficult to assess how much influence Cumming actually had on the design process, but much of it seems to have been attributable to the builder.

The four 'Clans' delivered in 1919 were very handsome machines and lived up to their design performance criteria. They were lighter than the 'Rivers' but had a higher tractive effort due to increased boiler pressure. The loading gauge of the Highland Railway was generous, but the 'Clans', like other classes from outside builders, had to be constructed to the inferior loading gauge of the North Eastern and Caledonian Railways over which they had to travel to reach Perth. Four further locomotives were delivered in 1921, the order being halved when the HR directors discovered the full cost.

The 'Clans' performed well throughout the system and were considered to be both efficient and reliable. They also had a very good power to weight ratio which made them extremely useful later on. The 'Clan' Class was Cumming's final design and also the final design for the Highland Railway, as the grouping of 1923 made it part of the London, Midland and Scottish Railway. When displaced from the Highland main line by the arrival of Stanier 'Black 5s', the 'Clans' were moved to the Oban line of the former Caledonian Railway. Weight restrictions on that line had always limited power potential but the 'Clans' were capable of hauling greater loads than the specially-designed Pickersgill 4-6-0s.

Withdrawals commenced in 1943 and the last survivor was scrapped in 1950. These robust and powerful locomotives did not get the acclaim of their brethren south of the border but they performed their work well and efficiently, and within the restrictions imposed they were certainly a very good design.

GWR 'CASTLE' CLASS 4-6-0

Designer C. B. Collett. **Introduced** 1923. **Power classification** 7P. **Driving wheel diameter** 6 ft 8.5 in. **Leading wheel diameter** 3 ft 2 in. **Cylinders** Four, 16 in diameter × 26 in stroke. **Valve gear** Walschaerts inside with rocking levers to outside cylinders. **Boiler** (Standard No 8 boiler as fitted from 1932 onwards)—*Pressure* 225 psi. *Grate area* 30 sq ft. **Heating surfaces**—*Tubes* 1,858 sq ft. *Firebox* 163 sq ft. *Superheater* 263 sq ft. **Tractive effort** 31,625 lbs. **Weight** Loco 79 tons 17 cwt, tender 46 tons 14 cwt. **Number built** 172 (including five rebuilt 'Stars' and the rebuilt 'Pacific' *The Great Bear*).

Churchward's 'Stars' were capable machines but by 1922 there was a growing need for an express locomotive capable of sustaining a higher power. Collett took the obvious course of action and further developed the already successful 'Star'. To have designed a completely new class might have satisfied the ego of an engineer at some other railway, but Collett was a good engineer and knew exactly what was needed. A proportional enlargement of the 'Star' was not possible without exceeding the axle weight restriction, so a completely new standard boiler was designed to supply steam to the slightly larger cylinders. Had it been possible to fit a larger steam generator, an even higher sustained maximum power could have been achieved. However, the 'Castles' were more powerful than the 'Stars' and due to better construction techniques were more efficient and more reliable.

Caerphilly Castle caused quite a stir when she was exhibited at the British Empire Exhibition held at Wembley in 1924, especially as the LNER 'Pacific *Flying Scotsman* was positioned alongside. The GWR publicity machine lost no opportunity to pronounce that the 'Castle' Class was the most powerful in Britain which, in terms of the dubious tractive effort criteria, it was. Friendly trials between the GWR 'Castle' Class and the LNER 'Pacifics' were arranged and they only served to show that the publicity was correct and further emphasized the greater economy of Collett's engine. In 1926, *Launceston Castle* ventured on to LMS metals and proceeded to demonstrate how inferior were the express locomotives of that Midland Railway dominated concern. Without doubt, Collett had produced a classic locomotive—it may have followed Churchward's original pattern, but it was Collett who had designed it.

In normal service, the 'Castles' were economical and fast. At one time, the GWR could claim to operate the world's fastest train,

Above *Didcot's preserved 'Castle', No 5051* Drysllwyn Castle.

Below *No 5005* Manorbier Castle *in streamlined form (GWR Museum, Swindon).*

BR-built Clun Castle *with double chimney.*

the 'Cheltenham Flyer'. Whilst operating that service in 1932, *Tregenna Castle* recorded the fastest ever time for the 77.3 miles between Swindon and Paddington, with a start-to-stop average speed of 81.6 mph.

Without much enthusiasm, Collett applied a grotesque form of streamlining to *Manorbier Castle* during the early months of 1935. Other railways were producing streamlined trains and some of the GWR hierarchy wanted to get in on the act, but the Chief Mechanical Engineer was not so easily impressed. His belief that normal steam locomotives were unsuitable for streamlining was ultimately proven correct by increased maintenance difficulties and lower than expected fuel savings.

Production of the 'Castles' continued until 1939 and more were built after the war and following nationalization. Although few would dispute the superiority of the design, deteriorating coal quality did cause problems and, in 1956, trials were carried out to see if the performance could be improved. The result was a modified draughting arrangement with a double chimney and a four-row higher-rate superheater. Even during the final years of steam in Britain, the 'Castles' were showing their paces and proving themselves a design that was one of the true classics of the steam age.

LMS 'JINTY' 0-6-0 TANK

Designer Henry Fowler. **Introduced** 1924. **Power classification** 3F. **Driving wheel diameter** 4 ft 7 in. **Cylinders** Two (inside) 18 in diameter × 26 in stroke. **Valve gear** Stephenson. **Boiler**—*Pressure* 160 psi. *Grate area* 16 sq ft. **Heating surfaces**—*Tubes* 967 sq ft (1,074 sq ft on some). *Firebox* 97 sq ft. **Tractive effort** 20,830 lbs. **Weight** 49 tons 10 cwt. **Number built** 422.

The unsung work-horse of any railway is its small shunting engine and for the LMS it was the 'Jinty'. That nickname, given by enthusiasts to this large class of 0-6-0 Class '3' tank locomotives, was actually first given to a group of 0-4-0 tank engines which worked the same railway.

After the grouping, some form of locomotive standardization was required within the LMS, and the higher management decreed that the Midland's 0-6-0 tank should be adopted for shunting. That design was not in fact new, but was based on a much earlier S. W. Johnson locomotive. Henry Fowler rebuilt a number of Johnson's machines incorporating a number of new features including a Belpaire boiler and an improved cab, but the same basic 0-6-0 tank design still remained.

George Hughes was actually Chief Mechanical Engineer of the LMS during its first two years of existence and it was then that the decision was taken to standardize on the 'Jinty'. In 1925, Fowler took command and continued production of his own design. There was, in reality, nothing special about the locomotive; it was simply tried and tested, and many of the earlier imperfections had long been eradicated. Longevity in a design has its advantages.

The 'Jinties' were of a straightforward design without fancy features such as superheating, the nature of their work not requiring it. For shunting and station pilot duties, a reliable and uncomplicated machine was essential, and the 'Jinties' were simply that. Most had vacuum brake equipment allowing them to operate on branch line freights, whilst some were also provided with steam heating to allow the haulage of passenger trains during cold weather. Whether on shunting or branch line duties, they proved to be capable and consistent performers. A number were taken into War Department stock and sent for service in France, and some were not returned to Britain until 1948 following a period of operation on French railways. The fact that conscription to WD duties came at all indicates how respected these little engines were.

'Jinty', BR No 47383, on Severn Valley Railway passenger duties at Highley.

Many enthusiasts neglected the 'Jinties' because they were engaged on less glamorous work, but they were always about at stations, in shunting yards and on branch lines. Glamour does not equate with importance or achievement, and these little 0-6-0 tanks were essential to LMS and BR operations. Even the spread of dieselization through the steady introduction of 0-6-0 diesel shunters could not get rid of them quickly, and the last of the class was not withdrawn until 1967, only a year before main-line steam disappeared from British Railways.

It may seem strange to advocate that a small, very basic 0-6-0 tank locomotive should be considered as a classic, but why not? They did their duty with efficiency and reliability, being more versatile than many of their larger sisters. Few express passenger designs lasted as long even though they may have looked more attractive. In any case, beauty is in the eye of the beholder, and there can have been few things more pleasurable than the sight of a 'Jinty' hauling a three-coach branch line train. It is pleasing that so many of these little engines have survived, allowing the Severn Valley Railway and other preserved lines to offer a repeat of that branch line scene.

SOUTHERN RAILWAY 'KING ARTHUR' CLASS 4-6-0

Designer R. E. L. Maunsell (after a Urie 1918 design). **Introduced** 1925. **Power classification** 5P. **Driving wheel diameter** 6 ft 7 in. **Leading wheel diameter** 3 ft 7 in. **Cylinders** Two (outside) 20.5 in diameter × 28 in stroke. **Valve gear** Walschaerts. **Boiler**—*Pressure* 200 psi. *Grate area* 30 sq ft. **Heating surfaces**—*Tubes* 1,716 sq ft. *Firebox* 162 sq ft. *Superheater* 337 sq ft. **Tractive effort** 25,320 lbs. **Weight** Loco 80 tons 19 cwt, tender 57 tons 11 cwt. **Number built** 74 (including 20 Urie locomotives).

During the first years following the grouping, the Southern had an unenviably bad reputation for almost everything, at least as far as the press was concerned. Its General Manager, Sir Herbert Walker, decided on image promotion and one aspect of this was the naming of express passenger locomotives. Maunsell, the

Preserved Southern Railway 'King Arthur' Class No 777 Sir Lamiel *with an enthusiasts' special at Appleby.*

CME, had no objections but warned that names would make no difference to the working of the engines. The first group selected was the newly-constructed 4-6-0 class built as replacements for the Drummond 'G14' engines. Names from Arthurian legend were chosen and hence the 'King Arthur' Class came into being.

The locomotives were claimed to be part of a new era on the new Southern, but in fact they differed very little from the Urie 'N15' Class built at Eastleigh in 1918. The 'N15s' were good runners and it was reasonable for Maunsell to upgrade the design to form a new class. The cylinder diameter was reduced from 22 in and long travel valves were introduced. The boiler pressure was raised by 20 lbs and the superheater area increased by 29 sq ft. Visual changes lay in the chimney shape, the provision of outside steam-pipes and a new cab design, but they were not overwhelming differences. Urie's 'N15s' were subsequently modified, given suitable names and incorporated into the class.

Ten Maunsell 'King Arthurs' were ordered and constructed at Eastleigh, as no other Southern works had experience of building anything larger than a 4-4-0. At the same time it became evident that the Traffic Department could not wait for development of the four-cylinder 4-6-0 'Lord Nelson' Class then being designed, as more power was required immediately for the central and eastern sections. Thirty 'King Arthurs' were therefore ordered, but Eastleigh could not cope and the North British Locomotive Co was asked to supply the machines. With the help of plans and moulds, that remarkable Scottish establishment commenced delivery less than six months after the order was placed, and completed the task within a year. A further 14 'Arthurs' were supplied by Eastleigh during 1926 and 1927.

The early promise of economical and fast running as demonstrated by No 451 *Sir Lamorak* was not sustained, especially on the south-eastern section. Some blame was put on inadequate crew training and some on poor track condition, but in truth it was a mixture of both. After the track had been upgraded and the crews taught a firing and driving routine which suited the locomotive, matters improved considerably.

Drifting smoke frequently obscured the driver's view, the first time such a problem had showed itself with a British locomotive. The fault was subsequently attributed to the use of the long lap, long travel valves when the engine was being driven at short cut-off. Great Western locomotives had similar arrangements but no drifting smoke problem, the taper boiler and relatively tall

BR 'King Arthur' No 30782 Sir Brian *at Bournemouth on 26 January 1956*
(D. K. Jones Collection).

chimney allowing the smoke to clear, but the short chimney of the
'Arthur' caused the smoke to cling. Several patterns of smoke
deflector were tried in extensive trials until the vertical plate type
demonstrated itself to be the most effective; this was the first
application of smoke deflectors to a British locomotive.

Considering that the design originated with Urie in 1918, the
'King Arthur' Class had a very long operational life; the last
examples were not withdrawn from traffic until 1962. Increasing
electrification of the Southern Region brought about their demise
but impending electrification also stifled steam locomotive
development. Had the Southern not adopted that policy, the
'Arthurs' might have been replaced, but then again they were such
useful and adaptable engines that they might well have lasted
much longer.

LNER 'U1' CLASS GARRATT
2-8-0+0-8-2T

Designer N. Gresley/H. W. Garratt/Beyer, Peacock & Co. **Introduced** 1925. **Driving wheel diameter** 4 ft 8 in. **Leading/trailing wheel diameter** 2 ft 8 in. **Cylinders** Six (arranged in two groups of one inside and two outside) 18.5 in diameter × 26 in stroke. **Valve gear** Walschaerts for outside cylinders, Gresley derived for inside cylinders. **Boiler**—*Pressure* 180 psi. *Grate area* 56.5 sq ft. **Heating surfaces**—*Tubes* 2,645 sq ft. *Firebox* 224 sq ft. *Superheater* 650 sq ft. **Tractive effort** 72,940 lbs. **Weight** 178 tons 1 cwt. **Number built** 1.

The Garratt type of articulated locomotive was the brain-child of Herbert William Garratt, a British engineer with experience in both the marine and railway industries. At first he could find no manufacturer interested in financing his idea, but eventually he succeeded in convincing Beyer, Peacock & Co that it was a sound investment. Garratt died in 1913, four years after the first locomotives to his design had been constructed for the railways of Tasmania; the original locomotive has now been re-imported and is held at the National Railway Museum, York. Beyer, Peacock & Co, however, went on to develop the design and produced large numbers of various powers and to various gauges for the railways of the world.

In Britain, the Garratt never really caught on as requirements for high-powered articulated locomotives were somewhat limited. Main-line applications were confined to 33 examples constructed by the LMS and a solitary engine for the LNER. That single locomotive was built for the purpose of banking on the 2.5 mile Worsborough incline between Wath and Penistone, with its gradient of 1 in 40. It was the most powerful locomotive ever constructed for duties in Britain and was said to be the equivalent of two 2-8-0 locomotives. Its use, therefore, saved one engine crew, but the duty of the single unlucky fireman was very arduous indeed.

The Great Central Railway had, in 1910, given some thought to construction of a Garratt type locomotive for banking duty, but nothing came of the idea until just after the grouping. Then, the LNER approached Beyer, Peacock & Co with a view to constructing two Garratts based on the Great Central Robinson-designed 2-8-0, but Gresley's preference for the three-cylinder arrangement soon gained favour, so the design that was finally

Above City of Truro *and* Drysllwyn Castle *in steam at Bridgnorth.*

Below *Preserved 'Duchess' 'Pacific'* Duchess of Hamilton *on the Settle-Carlisle line, heading south near Appleby.*

Above *Southern Railway 'King Arthur'* Sir Lamiel *working the Scarborough Spa Express in 1985.*

Below Lord Nelson *hauls its excursion train past Horton in Ribblesdale during August 1985.*

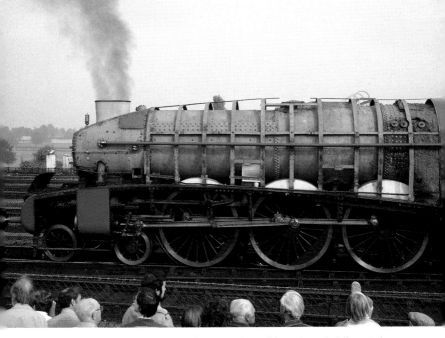

Above *Uncovered 'A4' –* Mallard *in steam but without her cladding at the National Railway Museum during restoration in 1985.*

Below Flying Scotsman *near Ais Gill with a southbound enthusiasts' train in August 1983.*

Above *'9F'* Evening Star *heads for Scarborough in August 1985.*

Below *The original Class '40' No D200 on duty at Southport in 1985.*

LNER 2-8-0+0-8-2T 'U1' Class Garratt No 2395, the most powerful locomotive to operate in Britain (National Railway Museum, York).

arrived at was based on the Gresley '02' Class 2-8-0 of the Great Northern. In basic terms, the machine consisted of two '02' engine units placed back-to-back with a boiler connecting them through pivots. It was built at the Gorton works of Beyer, Peacock & Co in 1925 and immediately replaced the two locomotives formerly required for the banking duties.

Throughout its long career on the Worsborough incline, the engine slogged its way up the gradient about 18 times in each 24 hour period. Availability was good, and only visits to the works for overhaul kept No 2395, later No 69999 under BR numbering, from its duties. Electrification of the Manchester-Sheffield-Wath lines over the Pennines in 1949 removed the need for banking, and the former LNER Garratt was moved by British Railways to the Lickey incline. There, the interloping LNER machine did not prove popular with LMS-trained men, and it was very soon returned to the Eastern Region.

The very heavy work required in firing the locomotive gave cause for concern, but the fitting of a mechanical stoker was considered inappropriate. Oil-firing offered the only salvation and a system was fitted in 1952. This proved to be moderately successful, resulting in a return to the Lickey at the beginning of 1955. However, the locomotive remained unpopular and, with the boiler life-expired by November 1955, a decision was made to withdraw and scrap.

As the most powerful steam locomotive ever to operate in Britain, this LNER Garratt deserves inclusion in a list of classic locomotives. It was designed and constructed for a specific purpose which it performed regularly and reliably for many years.

SOUTHERN RAILWAY 'LORD NELSON' CLASS 4-6-0

Designer R. E. L. Maunsell. **Introduced** 1926. **Power classification** 7P.
Driving wheel diameter 6 ft 7 in. **Leading wheel diameter** 3 ft 1 in.
Cylinders Four 16.5 in diameter × 26 in stroke. **Valve gear** Walschaerts.
Boiler (Typical values for the standard boiler fitted to most of the
class)—*Pressure* 220 psi. *Grate area* 33 sq ft. **Heating surfaces**—*Tubes*
1,795 sq ft. *Firebox* 194 sq ft. *Superheater* 376 sq ft. **Tractive effort**
33,510 lbs. **Weight** Loco 83 tons 10 cwt, tender 56 tons 14 cwt. **Number
built** 16.

The 'Lord Nelson' Class really owes its fame to what it might have
been rather than what it was. By virtue of the innovative features of
their design, the engines could be considered as classics, but
reality proved them to be something of a let-down; however, later
modifications helped a little with their tarnished image.

Whilst the 'King Arthur' Class was being developed, plans were
also being made for a more powerful passenger locomotive.
Harold Holcroft, an expatriate from the Great Western, held the
belief that on a four-cylinder locomotive the cranks should be set
at 135° rather than 180° in order to produce a more even torque
and blast. After limited trials on a Drummond 4-6-0, No E449, the
theory was accepted for inclusion in the new design. A wheel
diameter of 6 ft 7 in was decided upon for no better reason than it
was a stock size at Eastleigh works. The outside cylinders drove
the centre driving wheels, with cranks set at 90° to each other,
whilst the inside cylinders, with cranks also at 90° to each other but

Preserved Southern Railway 'Lord Nelson' Class No 850 Lord Nelson *at
York.*

advanced 45° relative to the outside cranks, drove the leading coupled wheels. In June 1925, an order for 15 'King Arthurs' was amended to 14 with one of the new type of 4-6-0; Maunsell followed his usual cautious approach and only constructed one member of this new class initially, thus allowing for complete appraisal before committing himself to full production.

On 27 September 1926, the prototype, *Lord Nelson*, made its public debut at Waterloo. Using the rather simplistic criteria of tractive effort, the Southern's Publicity Department made the most of the new machine as 'the most powerful Passenger Locomotive in Britain'. A single locomotive could not be run in traffic as part of a top link on an enhanced schedule, so the testing was not as rigorous as it might have been. Performance was not spectacular but was sufficiently good for the other 15 'Nelsons' to be ordered in two batches during 1928 and 1929. All the members of the class were given the names of famous British seafarers, and though most people had heard of such heroes as *Sir Francis Drake* and *Lord Rodney*, few seemed to know about the maritime connections of *Howard of Effingham* (he was the commander of the English fleet which defeated the Spanish Armada).

The class had been designed to meet a Traffic Department demand for a machine which could haul 500 tons at an average of 55 mph. Used on the 'Atlantic Coast Express' and 'Golden Arrow' boat trains, they proved capable but not outstanding. Maunsell attempted several modifications, particularly to the valves and exhaust, but it was not until O. V. S. Bulleid took command that real improvements came about. New cylinder arrangements and exhaust systems allowed the 'Nelsons' to achieve a performance close to that originally intended. The Lemaitre exhaust produced a change in the engines' appearance, with the original narrow chimney giving way to one of dominatingly wide proportions. As designed, the crank arrangements produced eight exhaust beats during each wheel revolution and the new blastpipe arrangement quietened the exhaust even further.

In power and performance terms, the class failed to set the locomotive world alight (a not inappropriate expression, as the soft exhaust blast did not pull the fire to pieces and thus avoided sparks which could indeed set the surrounding countryside alight!). They were, however, steady and reliable performers, providing valuable information for Maunsell in developing his 'Schools' Class and for Bulleid in the production of his 'Pacifics'. Fortunately, *Lord Nelson* himself has survived.

GWR 'KING' CLASS 4-6-0

Designer C. B. Collett. **Introduced** 1927. **Power classification** 8P.
Driving wheel diameter 6 ft 6 in. **Leading wheel diameter** 3 ft 0 in.
Cylinders Four, 16.25 in diameter × 28 in stroke. **Valve gear**
Walschaerts inside with rocking levers to outside cylinders. **Boiler**—
Pressure 250 psi. *Grate area* 34.3 sq ft. **Heating surfaces**—*Tubes*
2,007.5 sq ft. *Firebox* 193.5 sq ft. *Superheater* 313 sq ft. **Tractive effort**
40,290 lbs. **Weight** Loco 89 tons, tender 46 tons 14 cwt. **Number built**
30.

Passenger train loads during the middle years of the 1920s
increased steadily to the extent that the Great Western's 'Castles'
were pushed to their limits. A more powerful design was required
for the heaviest trains envisaged in the foreseeable future, but
after that no further increase in power would need to be
contemplated for the very good reason that platform lengths
imposed a limit on the size of any train. The number of new
high-powered locomotives required was, of necessity, small.

Felix Pole, then General Manager of the GWR, was also
interested in wresting the title of 'Most Powerful Locomotive' away
from the Southern's 'Lord Nelsons'; the basis of deciding power
merely on tractive effort was contentious, but prestige counted for
a lot with Pole.

Collett accepted the instruction and set about his design which
would be based upon the highly successful 'Castle' Class but built
to the limit of axle loading allowed by the Civil Engineer. This had
been raised to 22.5 tons on the main lines to the west as far as
Plymouth and north to Wolverhampton, and this restricted route
availability would keep the class very much confined. However,
Collett certainly produced an ace with his 'King' Class, even
though some are apt to decry the achievement by suggesting that
it was merely an enlargement of Churchward's 'Stars'. Basing the
new design on an established and successful one was good

'The King' – No 6000 King George V *with an enthusiasts' special train at
Craven Arms.*

engineering and economic sense and the 'Kings' spoke for themselves when the questions of power and performance were raised.

One of the class ventured across the Atlantic. No 6000, the first of the 'Kings' and named *King George V*, represented the GWR and Britain at the centenary of the Baltimore and Ohio Railway in 1927. For that performance, 'The King', as No 6000 is affectionately known, received two commemorative medals and a bell. The bell still graces the front buffer beam and the medals the cab side. However, a significant problem manifested itself whilst the engine was in America. Partial derailment of sister engine *King George IV* at Midgham was attributed to bogie springing and urgent rectification was necessary in case 'The King' similarly disgraced himself in front of the Americans.

Following the Second World War, there was a shortage of good quality Welsh steam coal for which the 'Kings' had been designed, and performance consequently suffered. However, modifications such as the fitting of a four-row superheater and a double chimney improved the steaming to an extent which surprised die-hard Great Western enthusiasts.

Weight limitations on the Great Western certainly restricted the 'Kings' prior to nationalization, and the ample GWR loading gauge to which they had been constructed prevented widespread use elsewhere after 1948. Despite that, the 'Kings' were classics in every sense. They were the most powerful 4-6-0s ever to run in Britain, with a style and grace to match. The exception was the ill-conceived and half-hearted attempt to streamline *King Henry VII*; Collett appears to have had little feel for the idea and the casing was removed as soon as possible.

Fortunately, three kings have been preserved; *King George V* is part of the National Collection whilst *King Edward I* and *King Edward II* have survived cutting up at Barry scrapyard to be restored with dedication and enthusiasm.

Kings from scrap – the rather dejected-looking King Edward II *at Bristol Temple Meads in 1985 awaiting restoration.*

LNER 'A3' CLASS 'PACIFIC' 4-6-2

Designer Nigel Gresley. **Introduced** 1927. **Power classification** 7P.
Driving wheel diameter 6 ft 8 in. **Leading wheel diameter** 3 ft 2 in.
Trailing wheel diameter 3 ft 8 in. **Cylinders** Three 19 in diameter × 26 in
stroke. **Valve gear** Walschaerts outside, derived motion for inside
cylinder. **Boiler**—*Pressure* 220 psi. *Grate area* 41.25 sq ft. **Heating
surfaces**—*Tubes* 2,476 sq ft. *Firebox* 215 sq ft. *Superheater* 635.5 sq ft.
Tractive effort 32,910 lbs. **Weight** Loco 96 tons 5 cwt, tender 58 tons.
Number built 78 (including 51 rebuilt 'A1' and 'A10' 'Pacifics').

Gresley's first 'Pacific' design, the 'A1', appeared in 1922 during
the final days of the Great Northern Railway. *Great Northern*, as
the initial locomotive was called, impressed everybody with its size
and power and a second machine quickly followed. An additional
ten 'A1s' were soon ordered but they entered traffic under the
LNER. First of that batch was No 4472 *Flying Scotsman*, which
found lasting fame in this country and abroad. A limiting aspect of
the design was the fact that they were dimensioned to suit the
ample GNR loading gauge which was somewhat more generous
than that of other regions of the new LNER. To allow working in
Northern and Scottish areas, a slight reduction in overall height
was required.

   Without doubt, the 'A1s' were good and a major improvement
on the large Ivatt and Raven locomotives which hauled the express
passenger trains on the East Coast main line. Exchange trials with
Collett's 'Castle' Class following the 1924 Empire Exhibition at
Wembley, however, soon showed that improvements were
possible, particularly with regard to the valve gear. Redesign of the
gear and the use of long travel valves showed an immediate
improvement in efficiency with the result that all 'A1s' were so
modified. During the mid-1920s, Gresley carried out superheater
experiments and decided that an improved higher pressure boiler
would be beneficial. Trials carried out with five 'A1s' illustrated the
new boiler's superiority and these 'Pacifics' were then designated
'A3'.

   The new boiler not only worked at a higher pressure but had an
increased superheat area which suited the Gresley concept,
although differing from the low temperature type which had served
the GWR so well at the exchange trials. 'Pacific' construction
between 1927 and 1935 followed the new 'A3' pattern, the only
other major difference being a reduction of one inch in the cylinder

'A3' 'Pacific' No 4472 Flying Scotsman *approaching Ais Gill summit in 1981.*

diameter. The former 'A1s' had their cylinders lined out to the new dimension or were fitted with new cylinder blocks when necessary; it was not until 1948 that the last 'A1' boiler conversion took place. The unofficial title 'super-Pacifics' was applied to the 'A3s' to distinguish them from the 'A1s'.

Some of the new engines were provided with corridor tenders in order that footplate crew exchanges could take place en route on the non-stop Scottish expresses. A banjo-shaped boiler dome instead of the circular dome was fitted to the final nine of the class and became a major distinguishing feature between the 'A1s' and 'A3s', as all new boilers were constructed to that form. The feature was not just cosmetic; it contained a slotted steam collector designed to prevent priming.

In 1937, Gresley fitted No 2751 *Humorist* with a double chimney and Kylchap blastpipe which improved steaming. As with many other railway matters, the war prevented any further development and it was not until twenty years later that Kenneth Cook, formerly of the GWR, applied the arrangement to the rest of the class. The 'A3s' had always been good performers on the fast express trains but by the 1950s their age was beginning to show although the Cook modification restored some of the former glory. Various smoke deflector forms were tried with the Kylchap blastpipes and eventually a German pattern 'wing' arrangement proved to be the most effective. Advancing dieselization brought the Gresley 'A3' era to an end but not before they had put up some outstanding performances alongside their replacements.

In a brief narrative it is impossible to fully detail the performances of these truly classic machines, and no attempt has been made to do so. The exploits of *Flying Scotsman* are almost legendary and, fortunately, still continuing, for that locomotive represents the class in preservation.

Final form for the 'A3s' with German-style smoke deflectors and double chimney. No 60036 Colombo *at Darlington, 8 August 1964* (D. K. Jones Collection).

GWR '5700' CLASS 0-6-0 TANK

Designer C. B. Collett. **Introduced** 1929. **Power classification** 3F. **Driving wheel diameter** 4 ft 7.5 in. **Cylinders** Two (inside) 17.5 in diameter × 24 in stroke. **Valve gear** Stephenson. **Boiler**—*Pressure* 200 psi. *Grate area* 15.3 sq ft. **Heating surfaces**—*Tubes* 1,076 sq ft. *Firebox* 102.3 sq ft. **Tractive effort** 22,512 lbs. **Weight** 47 tons 10 cwt. **Number built** 779.

There is no reason why the lowly shunting tank engine should not be a classic—that adjective is not restricted to express passenger locomotives. Indeed, the '5700' Class has more right to be so described than many others, for it served its owners with distinction over many years. Railways depended upon shunting engines to serve their main arteries, but the '5700' Class also operated branch line passenger and freight services and became an invaluable part of the Great Western's fleet. Axle loading restricted the class to GWR 'Blue' routes, a 16-18 ton limit. After 1950 they were allowed on to 'Yellow' routes with a 14-16 ton limit when the advantage of their negligible 'hammer blow' effect was recognized.

These pannier tanks were the natural successors to the numerous saddle and pannier tanks produced at Wolverhampton and Swindon around the turn of the century, and these earlier machines formed the basis for the '5700' design. So extensive was the class that it used up all numbers in the 57xx range and encroached upon the 67xx, 77xx, 87xx and 97xx ranges, then worked back through the 36xx, 37xx, 46xx and 96xx ranges! Not all the members of this class, the largest ever to operate on the GWR, were constructed at Swindon. The first batch actually came from the North British Locomotive Co whilst other contractors, including Beyer, Peacock & Co and Armstrong Whitworth, also supplied batches in different number ranges.

Not all the number ranges were identical in every respect as certain features were added or removed to suit particular circumstances. The 67xx engines, for instance, had steam brakes only, being intended solely for shunting purposes, whilst some 97xx machines were provided with condensing apparatus for working over the London Transport Metropolitan Line. Ease of maintenance, reliability and economy were required of shunting and branch line engines and few had any complaints about the Collett design on that account. The work was unglamorous but essential, and the '5700s' were the backbone of the system. If any

GWR '5700' Class pannier tank No 5764 at Bewdley on the Severn Valley Railway.

locomotive can be said to typify a railway, the '5700' pannier tanks and the Great Western Railway were synonymous.

Following nationalization, some '5700s' were used to replace similar types on the Southern Region, new construction at Swindon continuing into the British Railways era. The worth of these little machines was also obvious to those outside BR as a number were purchased for use by the National Coal Board and London Transport. At the mines they performed the usual shunting duties, while with LT they were employed on engineers' works trains, where service continued after their BR sisters had been consigned to the scrapyard.

As might be expected from such a large and useful class, 16 have been preserved, five coming from London Transport. Their versatility provides valuable support for the preserved railways of Britain, and appropriately the '5700' pannier tank has been immortalized as 'Duck' in the Rev Awdry's Railway Stories.

BR-built '5700' Class No 9681 on the Dean Forest Railway in 1986.

SOUTHERN RAILWAY 'SCHOOLS' CLASS 4-4-0

Designer R. E. L. Maunsell. **Introduced** 1930. **Power classification** 5P.
Driving wheel diameter 6 ft 7 in. **Leading wheel diameter** 3 ft 1 in.
Cylinders Three 16.5 in diameter × 26 in stroke. **Valve gear**
Walschaerts. **Boiler**—*Pressure* 220 psi. *Grate area* 28.3 sq ft. **Heating
surfaces**—*Tubes* 1,604 sq ft. *Firebox* 162 sq ft. *Superheater* 283 sq ft.
Tractive effort 25,130 lbs. **Weight** Loco 67 tons 2 cwt, tender 42 tons 8
cwt. **Number built** 40.

Having satisfied the Southern Railway's Traffic Department with
his 'Lord Nelson' Class, Maunsell found himself almost immediately faced with another motive power request. This time the
need was for a locomotive capable of hauling 400-ton trains at an
average start-to-stop speed of 55 mph, but with an axle loading of
less than 21 tons.

Improving existing lightweight designs quickly proved itself to
be of no use, so Maunsell turned to the idea of scaling down his
'Nelson' design. Removal of one inside cylinder and one driven
axle together with a shortened boiler seemed to be the solution.
The weight, however, was still above the limit, so the Belpaire
firebox, favoured by Maunsell, had to go. A 'King Arthur'-type
boiler with a round-topped firebox was adapted to suit the new
4-4-0 locomotive. Surprisingly, a feature incorporated in the
original Belpaire firebox proposal remained in the final design; to
overcome restricted clearances on the Tonbridge-Hastings route,
the cab sides had been raked inwards, but, with the adoption of the
round-topped firebox, a narrower cab could have been used
making this feature unnecessary. It remained, however, a
distinctive part of the 'Schools' design.

In most respects, the design of the 4-4-0 did not present major
problems as many of the features were taken from existing
classes. That did not, however, make the locomotive a compromp-
ise which satisfied nobody, and its performance was soon
recognized as being first rate. The first ten members of the class
appeared in 1930 and a decision was made to name them after
major schools. No naming ceremonies took place, but each
locomotive was put on display as near as possible to the school
after which it was named. For later locomotives this was not
possible, some schools being well outside the Southern operating
area. This naming was considered an honour by all concerned

Southern Railway 'Schools' Class No 30929 Malvern *at Ramsgate, 14 May 1956* (D. K. Jones Collection).

except in one instance. No 923 was named *Uppingham* but the headmaster of that school took offence for some strange reason and demanded that the name of his establishment be removed. This the SR acceded to and promptly renamed the locomotive *Bradfield*.

Early operations were as required, but by the middle of 1931 the Tunbridge Wells – Hastings line had been upgraded and the class was allocated to the route for which it had been originally intended. It soon became evident that the locomotives were a success, with an ability to haul heavier-than-expected trains whilst still being capable of making up time. The Southern's management had no second thoughts regarding the ordering of an additional twenty 'Schools'.

Gradually, these locomotives took up duties throughout the eastern section of the railway, being used for an assortment of train services with considerable success. An order for a further ten 'Schools' brought the class total to 40, but there were 41 sets of name-plates because of the Uppingham incident; that school actually had one of the plates its headmaster so vehemently demanded be removed!

The entire class performed invaluable service on Southern metals until the extension of electrification brought about steam's rapid demise during the early 1960s. The 'Schools' were the most powerful 4-4-0 locomotives ever to run in Britain and it is fitting that three of this rather attractive class should survive. *Repton* went to the USA following withdrawal but *Stowe* can be seen operating trains deep in the heart of Southern territory on the Bluebell Railway. *Cheltenham* forms part of the National Collection and normally resides at the National Railway Museum, York.

LMS 'PRINCESS ROYAL' CLASS 'PACIFIC' 4-6-2

Designer William A. Stanier. **Introduced** 1933. **Power classification** 8P. **Driving wheel diameter** 6 ft 6 in. **Leading wheel diameter** 3 ft 0 in. **Trailing wheel diameter** 3 ft 9 in. **Cylinders** Four, 16.25 in diameter × 28 in stroke. **Valve gear** Walschaerts. **Boiler**—*Pressure* 250 psi. *Grate area* 45 sq ft. **Heating surfaces**—*Tubes* 2,299 sq ft. *Firebox* 217 sq ft. *Superheater* 598 sq ft. **Tractive effort** 40,300 lbs. **Weight** Loco 104 tons 10 cwt, tender 54 tons 13 cwt. **Number built** 12.

For the 1933 schedules, the LMS management wanted to operate 500-ton trains on the Euston-Glasgow service, so a more powerful locomotive than the 'Royal Scots' would be needed. The new service also called for through working rather than a change of locomotive at Carlisle. William Stanier had but recently taken over as Chief Mechanical Engineer when, in July 1932, the instruction went out to prepare such a machine.

Within 18 months, the first of the class was completed at Crewe; it was a 'Pacific', No 6200, and its name designated the entire class, *Princess Royal*. Before the year's end, No 6201 *Princess Elizabeth* appeared, but the third locomotive approved as part of the initial batch did not; Stanier had plans for a turbine drive and No 6202 did not enter service until 1935. The 'Turbomotive', as it became known, was a bold adventure into an alternative steam drive using a turbine supplied by Metropolitan Vickers. Being a one-off, the 'Turbomotive' could not form part of a conventional link and was employed mainly on the Euston-Liverpool run. While the experiment was not a failure, it was not successful enough to warrant production of other turbine locomotives. When renewal of the turbine and other parts became necessary after nationalization, No 46202 (her BR number) was rebuilt in conventional form. *Princess Anne*, as the engine became, was damaged beyond repair in the 1952 Harrow disaster, barely three months after re-entering service.

Stanier's 'Princess Royal' design showed his Swindon upbringing in the domeless taper boiler and low degree of superheat. It soon became evident that such features did not suit LMS conditions and by the time the second batch of locomotives appeared in 1935, the boilers had been provided with domes and the superheat temperature increased. Stanier did break away from Swindon tradition by providing the footplate crew with a spacious

'Princess Royal' 'Pacific' Princess Louise *at Carlisle in 1962* (D. K. Jones Collection).

cab having tip-up seats and a folding window which acted as a windshield.

In service, the 'Princess Royals' were good and, by the time the initial problems had been rectified and the crews had become familiar with the firing and driving requirements, the accelerated service to Scotland became no problem. They could climb the hills at Shap and Beattock with ease, even with the heaviest of trains, and had a turn of speed to match any schedule given to them. On November 16 1936, *Princess Elizabeth* with a seven coach train of 225 tons ran non-stop between Euston and Glasgow in 354 minutes at an average speed of 68.1 mph. The return next day was accomplished in 344 minutes, at an average speed in excess of 70 mph, despite an extra coach being added, and these runs represented a world record for non-stop long-distance steam traction.

Success of the 'Princess Royals' resulted in the development of the 'Duchess' 'Pacifics' which then took over the main Glasgow express trains, but that is another story. Without a doubt, this Stanier 'Pacific' design elevated the LMS locomotive stock to a position it had never before held. The record runs of 1936 were invaluable in publicity terms, but everyday performance allowed the timings to be kept and that was even more important. The engines were powerful and durable with an ability to make steam under the most difficult of conditions. Like all locomotives, however, they had to be treated properly, but such was their popularity that shed and footplate staff took pride in working with them.

Beauty is subjective and a personal view, but the 'Princess Royal' Class was certainly one of the most attractive ever produced in Britain, and it also performed to expectation. With Nos 6201 and 6203 preserved, the reader can decide upon the attraction aspect for himself. Unfortunately, there is no longer the opportunity to demonstrate their full capabilities.

Preserved 'Princess Royal' Class No 6201 Princess Elizabeth *at Shrewsbury ready to haul a special train.*

LMS 'BLACK FIVE' 4-6-0

Designer William A. Stanier. **Introduced** 1934. **Power classification** 5MT. **Driving wheel diameter** 6 ft 0 in. **Leading wheel diameter** 3 ft 3.5 in. **Cylinders** Two (outside) 18.5 in diameter × 28 in stroke. **Valve gear** Walschaerts. **Boiler** (Typical values for domed boilers generally fitted)—*Pressure* 225 psi. *Grate area* 28.65 sq ft. **Heating surfaces**— *Tubes* 1,478.7 sq ft. *Firebox* 171.3 sq ft. *Superheater* 359.3 sq ft. **Tractive effort** 25,455 lbs. **Weight** Loco 70 tons 12 cwt, tender 54 tons 13 cwt. **Number built** 842.

When William Stanier joined the LMS in 1932, the locomotive situation on that railway was critical, with a powerful mixed-traffic locomotive being high on the list of priorities. The design staff was set to work to produce a machine which could operate virtually throughout the system, being at home both on express passenger trains and branch line freight. Subsequent events showed that they succeeded admirably. The 'Black Stanier', or 'Black Five' as it is now affectionately known, was able to replace numerous pre-grouping locomotives of various types. In view of the large number of engines initially required, and the fact that the LMS works were fully engaged, orders had to be placed with two outside contractors, the Vulcan Foundry and Armstrong Whitworth. Embarrassingly, Vulcan delivered No 5020 before Crewe could complete No 5000, nominally the first member of the class.

In common with other early Stanier designs, low-superheat domeless boilers were provided for the original batches. However, the Swindon practice upon which Stanier had been reared suited good quality Welsh coal which was not used on the LMS, so

The original 'Black Five' in number if not in construction. No 5000 at Arley on the Severn Valley Railway.

design changes quickly came about. Although slight variations did occur, the majority of the boilers were of the dimensions given above; some locomotives, however, retained their domeless boilers throughout their operating lives.

Popular with footplate crews and shed staff alike, the 'Black Fives' were to be seen throughout the LMS system from Wick in the north to Bournemouth on the south coast, equally at home hauling passengers or freight. Even when substituting for 'Pacifics' on express passenger duties, they could keep time, if in reasonable condition, having been designed for a maximum speed of 90 mph. They were efficient and fairly easy to maintain compared with similarly-powered machines of the period. It is not surprising, therefore, that steady production continued beyond nationalization. 'Black Fives' were delivered each year from 1934 until 1951, the LMS works at Crewe, Derby and Horwich supplying all the locomotives following the earlier order to contractors.

So reliable and well understood were the 'Black Fives' that H. G. Ivatt, final Chief Mechanical Engineer of the LMS, decided to use the basic design for experimenting. He developed batches having various combinations of Timken roller bearings, double blastpipes, outside Stephenson valve gear and Caprotti poppet valve gear. In all, a staggering 842 locomotives of this class were produced, a testimony to the perfection of its design. The definition of a classic as 'a work of recognized excellence' must surely apply here—had they not been so effective in performing all their duties, their numbers would have been much lower.

Beauty is a subjective characteristic which cannot always be readily applied to a machine, but Stanier's Class '5' mixed-traffic

'Black Five' with Stephenson link motion valve gear – No 4767, now named George Stephenson.

The classic lines of a 'Black Five', No 44932 at Southport.

4-6-0s were well-proportioned and looked attractive at the head of any train, whether express passenger or mixed freight. Whether or not they were beautiful is for the individual to decide. Even the grime of neglect during the final years of British steam could not hide the clear aesthetic lines of the 'Black Five'; if not beauty, they certainly had style.

Many features, including the boiler, were subsequently utilized for the British Railways standard Class '5' 4-6-0 which was introduced in 1951. Indeed, the Caprotti valve version of the 'Black Five' pointed the way to the highly successful Caprotti valved standard Class '5'.

The fact that the last British main-line steam operations were on former LMS lines in the north-west of England certainly helped Stanier's 'black 'uns' to survive. It is, however, no accident that they lasted to the end of steam whilst their younger BR standard sisters preceded them to the scrapyard. The final scheduled steam passenger workings in 1968 were 'Black Five' hauled—only the best lasted until the end!

Fortunately, preservation has looked kindly upon the 'Black Fives' and many have escaped the cutter's torch. These include No 5000, nominally the first of the class, and No 44767, now named *George Stephenson* which was fitted with outside Stephenson valve gear. That so many were purchased directly from British Railways upon withdrawal illustrates the affection with which they were, and are, held. The sharp exhaust bark and deep whistle of Stanier's classic mixed-traffic locomotive may still be heard on main-line excursions and at preserved railways throughout Britain.

LMS '8F' CLASS 2-8-0

Designer William A. Stanier. **Introduced** 1935. **Power classification** 8F. **Driving wheel diameter** 4 ft 8.5 in. **Leading wheel diameter** 3 ft 3.5 in. **Cylinders** Two (outside) 18.5 in diameter × 28 in stroke. **Valve gear** Walschaerts. **Boiler**—*Pressure* 225 psi. *Grate area* 28.65 sq ft. **Heating surfaces**—*Tubes* 1,479 sq ft. *Firebox* 171 sq ft. *Superheater* 245 sq ft. **Tractive effort** 32,440 lbs. **Weight** Loco 70 tons 10 cwt, tender 54 tons 13 cwt. **Number built** 852.

Although the LMS had a stock of 2-8-0 and 0-8-0 heavy freight locomotives, they were not considered satisfactory for the entire system, especially with the loads experienced in the early 1930s. As with the passenger stock, Stanier provided the answer and produced a locomotive of outstanding quality.

In 1934, two experimental 2-8-0s of Stanier design were ordered in place of five Fowler machines of similar wheel arrangement, the idea being to gain experience before a larger order was placed. In the event, the order was increased to twelve with delivery taking place during 1935. As with other Stanier designs of the period, these locomotives, initially classified 7F, were given domeless taper boilers. Apart from one example, No 8003, these initial machines never received the later, more acceptable, domed boiler with top feed.

Within a short time, the worth of the design became evident and other orders followed; the domed and enlarged boiler was fitted to these, bringing the classification up to 8F. Vacuum braking equipment was also provided. With Crewe works unable to undertake construction at the required rate, the batch comprising Nos 8027-8095 was allocated to the Vulcan Foundry for delivery during 1936-37.

The outbreak of the Second World War produced a need for large numbers of heavy freight locomotives to operate overseas, and the War Department immediately saw the potential of Stanier's 8F, its power and reliability having been proven throughout the LMS system. The WD placed an order for locomotives to service the British expeditionary force in Europe, but the events which culminated in the evacuation at Dunkirk changed that. The theatre of war changed to the Middle East, with its associated logistic problems. There was also a pressing need to keep the Russian allies supplied, one route being via the Iranian railway. Powerful locomotives were required and the 8F was the obvious choice, so throughout the war years batches were constructed by the North

British Locomotive Co and Beyer, Peacock & Co as well as at the works of the 'big four' railway companies.

A number of 8Fs were taken into LNER stock and Hawksworth, on the Great Western, even considered the use of the 8F boiler for certain of his designs. Those locomotives that were sent abroad operated throughout the Middle East, especially in Egypt and Iran, and performed well in the service of Britain. At the end of hostilities, many were repatriated, but some stayed, victims of the conflict or still of use to the forces remaining. A number were retained by the railways of that region, several still being active on Turkish railways until the beginning of the 1980s. In 1986, the association of the class with the military was marked by the dedication of No 8233, now preserved on the Severn Valley Railway but formerly having served in Iran and Egypt, to the memory of British military railwaymen who died during the Second World War.

To have received such widespread use by the forces, the 8F must have been something special. Indeed it was, for in Britain members of the class saw out main-line steam, dominating traffic in Lancashire alongside the 'Black Fives' until the final workings in August 1968.

'8F' No 48431 rests at Haworth shed on the Worth Valley Railway.

LNER 'A4' CLASS 'PACIFIC' 4-6-2

Designer Nigel Gresley. **Introduced** 1935. **Power classification** 8P. **Driving wheel diameter** 6 ft 8 in. **Leading wheel diameter** 3 ft 2 in. **Trailing wheel diameter** 3 ft 8 in. **Cylinders** Three 18.5 in diameter × 26 in stroke. **Valve gear** Walschaerts with derived drive for inside cylinder. **Boiler** (Typical values for LNER days)—*Pressure* 250 psi. *Grate area* 41.25 sq ft. **Heating surfaces**—*Tubes* 2,345 sq ft. *Firebox* 231 sq ft. *Superheater* 749 sq ft. **Tractive effort** 35,455 lbs. **Weight** Loco 102 tons 19 cwt, tender 64 tons 3 cwt. **Number built** 34.

The 1932 performance of two-car diesel-electric sets in Germany so impressed the LNER directors that they enquired of their Chief Mechanical Engineer if he could discover more about their operations. This streamlined train, known as the 'Flying Hamburger', ran between Berlin and Hamburg and had a top speed of 100 mph. Gresley soon discovered that a similar three-coach train could operate the London to Newcastle service in 4.5 hours but with only 140 passengers transported in very basic comfort. That was unacceptable to the General Manager who encouraged his CME to run a trial using his 'A3' locomotives; the results showed that a high-speed steam service could be operated with more passengers in greater comfort.

A powerful locomotive was required and Gresley realized that streamlining would be ideal. During the design process, extensive wind tunnel tests were carried out at the National Physical Laboratory and, only 25 weeks after approval was obtained, the first 'A4' 'Pacific' appeared. The streamlined *Silver Link*, with its special coaches for the 'Silver Jubilee' train, was impressive. It was totally unlike any other locomotive on the rails of Britain, but some enthusiasts did not take kindly to the new shape. Nonetheless, the wedged-shape nosed and aerofoil running-plate over the wheels gave an impression of speed, and actual performance matched up to appearance; sustained speeds of up to 90 mph were called for in the 'Silver Jubilee' timings.

The initial four locomotives were so successful that a further batch was ordered as part of the 1936 renewal programme, with another 14 coming immediately afterwards. Four of the final group were given Kylchap exhaust systems and large double chimneys which not only looked better but improved steaming. Only in BR days were the other members of the class similarly treated. Most of the 'A4s' were fitted with corridor tenders in order that footplate crew changes might take place without the need for a stop en route.

One of the locomotives given the Kylchap exhaust from new was No 4468 *Mallard* which proceeded to set a world speed record for steam of 126 mph during high-speed braking trials on July 3 1938—that record still stands.

Wartime conditions produced a change in shape—the valances were removed from in front of the cylinders and over the coupled wheels in order to expose those parts for easier servicing, and were never replaced. The chime whistle remained, however, a feature which made the approach of an 'A4' obvious even when it was still out of sight. Poor quality coal during the post-war period resulted in indifferent steaming and after nationalization all the members of the class were provided with modified exhaust systems and double chimneys. Attention was also given to the valve gear, as the derived motion for the inside cylinder became defective due to pin and bush wear. These modifications restored the class to pre-war performance levels, although train timings did

Gresley's streamlined 'A4' 'Pacific' Sir Nigel Gresley *– the view from above shows the double chimney.*

'A4' in BR livery – No 60009 Union of South Africa *at Dundee, 9 May 1981.*

not warrant such high-speed running on all services.

No other British attempt at streamlining was so successful as that applied to the 'A4s'. How effective it was in reducing coal consumption during normal service is a matter for speculation but, apart from the removal of the valances to improve accessibility for maintenance, the entire class remained streamlined. It is an interesting mental exercise to imagine the shape of a de-streamlined 'A4', and who is to say it would have been less attractive—Stanier's 'Duchess' class improved in appearance when the streamlining was removed.

Without a doubt, the 'A4s' were classics, not only because of their performance and *Mallard*'s record but because of the era they introduced. The high-speed running of the 1930s is a part of their history, and though the shape may have upset some purists when first put on show, few would now disagree with the generally held view that the double-chimneyed 'A4s' had style.

Six of the class have been preserved, including the record breaking *Mallard*. Two are overseas, *Dominion of Canada* in that country and *Dwight D. Eisenhower*, appropriately, in the USA.

LNER 'V2' CLASS 2-6-2

Designer Nigel Gresley. **Introduced** 1936. **Power classification** 6MT. **Driving wheel diameter** 6 ft 2 in. **Leading wheel diameter** 3 ft 2 in. **Trailing wheel diameter** 3 ft 8 in. **Cylinders** Three 18.5 in diameter × 26 in stroke. **Valve gear** Walschaerts outside with derived motion for the inside cylinder. **Boiler**—*Pressure* 220 psi. *Grate area* 41.25 sq ft. **Heating surfaces**—*Tubes* 2,216 sq ft. *Firebox* 215 sq ft. *Superheater* 680 sq ft. **Tractive effort** 33,730 lbs. **Weight** Loco 93 tons 2 cwt, tender 51 tons. **Number built** 184.

Gresley's 'Pacifics' gave the LNER ample motive power for all the passenger trains of the mid-1930s, but freight traffic was meeting serious competition from road haulage. Management decided to introduce a fast parcels service under the name 'Green Arrow', and a new locomotive would be required as the 'Pacifics' were needed for other services and the existing powerful 2-6-0s, the 'K3s', were considered unsuitable. Eventually Gresley settled on a 2-6-2 wheel arrangement, and the 'V2' Class was born.

The boiler was a shortened version of that applied to the 'A3' 'Pacifics', giving the 'V2' a similar profile. The coupled wheelbase was, however, larger than that of the 'Pacifics', the smaller-diameter driving wheels being spaced further apart. These wheels were larger than those normally applied to mixed-traffic locomotives and that allowed the 'V2s' to substitute for 'Pacifics' on express passenger trains.

Four members of the class were produced at Doncaster in 1936, the first appropriately being named *Green Arrow*, and they were allocated to different areas so as to provide experience under varied operating conditions. In terms of power, performance and reliability they could not be faulted, at least whilst still relatively new, living up to the usual high Gresley standards. Axle loading was higher than that for similar mixed-traffic locomotives such as the LMS 'Black Fives' and the GWR 'Halls', and that restricted their route availability to main lines. The 'V2s' were, however, power class '6' rather than class '5'.

Whilst the conjugated valve gear remained in good condition, their performance could match the 'A3' Pacifics, but with the onset of wear the valve settings suffered with a consequent fall-off in performance. The initial four locomotives proved to be so effective at the job for which they were designed that a further batch was ordered for delivery the following year, and production then continued at a steady rate until 1944. That wartime production contradicted the accepted dictate which only allowed production of heavy freight locomotives, for the overall usefulness of the class ensured that production during those war years would play a major part in the war effort; some people consider the 'V2s' to be the locomotives which helped win the war. Certainly their versatility was extremely advantageous in that they could be utilized for a variety of services from heavily-loaded passenger trains to the even heavier cross country freights.

After post-war services had returned to normal by the mid-1950s, the 'V2s' were once again being utilized for passenger and freight services with performances equal to those of the pre-war period, again provided that they were well maintained with valve gear in good condition. An experiment of fitting a Kylchap double blastpipe showed considerable merit and five members of the class were so treated. Unfortunately, the spread of dieselization brought the experiment, and the 'V2s' themselves, to an end.

Green Arrow now forms part of the National Collection and makes regular forays on to the main line with excursions. Such trains can be managed with ease, indicating little of the power that the 'V2s' exerted whilst hauling a variety of trains during the dark days of the Second World War. For this high-powered mixed-traffic class, Gresley used many features from his other designs which blended in such a way as to produce a locomotive of great versatility, the need for which became patently obvious during those war years. The country was well served by these machines.

'V2' Class No 60859 at Darlington on 8 July 1963 (D. K. Jones Collection).

LMS 'PRINCESS CORONATION' OR 'DUCHESS' CLASS 'PACIFIC' 4-6-2

Designer William A. Stanier. **Introduced** 1937. **Power classification** 8P. **Driving wheel diameter** 6 ft 9 in. **Leading wheel diameter** 3 ft 0 in. **Trailing wheel diameter** 3 ft 9 in. **Cylinders** Four, 16.5 in diameter × 28 in stroke. **Valve gear** Walschaerts. **Boiler**—*Pressure* 250 psi. *Grate area* 50 sq ft. **Heating surfaces**—*Tubes* 2,577 sq ft. *Firebox* 230 sq ft. *Superheater* 856 sq ft (Later reduced to 830 sq ft). **Tractive effort** 40,000 lbs. **Weight** Loco (streamlined) 108 tons 2 cwt, loco (non-streamlined) 105 tons 5 cwt, tender 56 tons 7 cwt. **Number built** 38.

The 'Princess Royal' 'Pacifics' had performed well on the express train services between London and Scotland but by 1936 it was realized that an improved locomotive was required if faster timings were to be achieved. Work on a new design was set in hand but Stanier had little to do with the detail as he was in India at the time.

Streamlined LMS 'Duchess' 'Pacific' No 6221 Queen Elizabeth *passing South Kenton with the 'Coronation Scot' in 1937* (National Railway Museum, York).

Preserved 'Duchess' No 46229 Duchess of Hamilton.

Tom Coleman, Senior Draughtsman at Derby, was entrusted with the work.

For high-speed running, the driving wheel diameter was increased by 3 inches compared with the earlier 'Pacifics' and, in order to maintain tractive effort, the cylinder diameter was also increased. To feed these enlarged cylinders, a bigger boiler had to be provided. Low-degree superheat as favoured for the original 'Princess Royal' design was abandoned, and the new machines were fitted with the largest superheater of any British locomotive. For simplicity, only two sets of valve gear were provided, the inside cylinders having valves operated by rocking levers from the outside valve spindle crossheads.

At the time, streamlining was very much in vogue and the directors agreed that publicity would be gained by introducing a streamlined train to match that of the LNER. The aerodynamic casing fitted to the original batch of the 'Duchess' 'Pacifics' actually increased the weight by some two tons compared with the basic design, and how effective that casing was in reducing wind resistance in practice is difficult to determine, but it was eye-catching and created an impression of speed. The locomotives themselves lived up to that impression and the new 'Coronation Scot' train service offered a 6.5 hour journey between London and Glasgow. Only caution on the part of the Running Department prevented a faster service; that the locomotives were capable of even higher speeds was proved by the first member of the class No 6220 *Coronation* when, in June 1937, she set a world speed record for steam of 114 mph.

The first batch of engines steamed and rode very well, indicating that the design was an absolute success. More of the streamlined locomotives were soon ordered for operating all the main express services to Liverpool and Scotland, but these appeared in a gilt-lined maroon livery rather than the blue of the special 'Coronation Scot' locomotives. Enthusiasm for streamlining cooled somewhat as the added problems for maintenance and servicing became evident, and the next batch of 'Duchess' 'Pacifics' appeared without the streamlined casing. They were identical in every other respect but were lighter and very attractively proportioned. Up until then, the casing had hidden the graceful lines of what was, underneath, a conventional steam locomotive. In later years, the streamlined engines had their casings removed, the only constructional change required being to the smokebox which had a downward taper to allow the fitting of the casing. Most of the de-streamlined engines, however, ran with these smokeboxes until they required renewal.

In streamlined or conventional form, the 'Duchess' 'Pacifics' were first-class locomotives. It is doubtful if they ever achieved their maximum power for more than short periods of time due to the limits imposed on steam generation by manual firing. The fitting of a mechanical stoker would certainly have allowed the boiler to steam consistently to its full potential, but outbreak of war prevented further development of the 'Duchess' Class and the introduction of the standard 'Pacifics' following nationalization resulted in a loss of interest. Early withdrawal of the complete class between 1962 and 1964 left many enthusiasts with the feeling that there had been a deliberate policy to discriminate against these fine locomotives, especially as during their final years they had been demoted to secondary and parcels traffic. Arguably the best British 'Pacific' design, the 'Duchess' Class is represented in preservation by three examples, with *Duchess of Hamilton* being much in demand for excursion trains.

SOUTHERN RAILWAY 'Q1' CLASS 0-6-0

Designer O. V. S. Bulleid. **Introduced** 1942. **Power classification** 5F. **Driving wheel diameter** 5 ft 1 in. **Cylinders** Two (inside) 19 in diameter × 26 in stroke. **Valve gear** Stephenson. **Boiler**—*Pressure* 230 psi. *Grate area* 27 sq ft. **Heating surfaces**—*Tubes* 1,470 sq ft. *Firebox* 170 sq ft. *Superheater* 218 sq ft. **Tractive effort** 30,080 lbs. **Weight** Loco 51 tons 5 cwt, tender 38 tons 0 cwt. **Number built** 40.

The word 'ugly' was liberally applied in descriptions of the 'Q1' Class when it was introduced on the Southern Railway at the height of the war in 1942. Upon seeing a photograph of the machine for the first time, William Stanier is said to have commented 'I don't believe it' and then asked, 'Where's the key?', considering that it looked like a tin-plate toy. Certainly the style was unconventional and Bulleid paid no heed to aesthetics in producing the design. The midst of a World War was no time to be bothered with frills, and his machine had to be plain and functional, a true 'austerity' type; what it lacked in looks it certainly made up for in performance.

Bulleid's brief was for a powerful 0-6-0 which could operate over the same secondary routes as the earlier Maunsell 'Q' Class. An easy option would have been to construct further 'Q' Class engines, but Bulleid knew he could improve on the power whilst at the same time increase the availability and reduce maintenance. To do so he went back to basic principles and incorporated only essential features. His approach worked admirably.

Bulleid's strange-looking 0-6-0 'Q1' Class No 33021 (D. K. Jones Collection).

A 'Q1' from the rear – No 33004 at Guildford in 1964 (D. K. Jones Collection).

Lightweight cast steel wheels to the Bulleid-Firth-Brown patent, as used for the 'Merchant Navy' Class, were employed. The traditional running-boards above the wheels were dispensed with to save weight and simplify construction—they were totally unnecessary to the design anyway. It was the boiler which produced the most visually disturbing impact, but again Bulleid only included the absolute essentials. A simple frame supported the sheeting which enclosed the fibreglass lagging, that sheeting being in two distinct sections for convenience. The smokebox had a flat bottom in order that it might sit easily on the frame and cylinder casting, and the chimney and dome were similarly basic.

Not only was the 'Q1' more powerful than the 'Q' Class but it also exceeded the power of comparable 0-6-0s on the LMS and LNER. At the same time, it could be compared favourably with LMS 2-8-0s and Riddles 'Austerity' 2-8-0s in terms of tractive effort per pound weight and per pound cost. Formally limited to 55 mph, the 'Q1s' could operate safely and conveniently at up to 75 mph, even tender first, although there was a tendency to roll at the higher speeds.

Not a real contender in the beauty stakes, the 'Q1s' were notable for the fact that they did exactly what was asked of them. Bulleid's design fitted the prevailing circumstances and seemingly satisfied everybody. The operating department was satisfied with the high power and availability, footplate crews liked the ease of working and relative comfort, maintenance staff benefitted from the accessibility and the company was happy with the low initial cost. Few locomotives can have pleased so many people, surely a sign of excellence.

Only the observer appears to have been dissatisfied with the 'Q1', and that because of its styling. A single example has been preserved, No 33001, on the Bluebell Railway, and with the passage of time the profile begins to look less harsh, even ruggedly attractive.

LNER 'B1' CLASS 4-6-0

Designer Edward Thompson. **Introduced** 1942. **Power classification** 5MT. **Driving wheel diameter** 6 ft 2 in. **Leading wheel diameter** 3 ft 2 in. **Cylinders** Two (outside) 20 in diameter × 26 in stroke. **Valve gear** Walschaerts. **Boiler**—*Pressure* 225 psi. *Grate area* 27.9 sq ft. **Heating surfaces**—*Tubes* 1,493 sq ft. *Firebox* 168 sq ft. *Superheater* 344 sq ft. **Tractive effort** 26,880 lbs. **Weight** Loco 71 tons 3 cwt, tender 52 tons. **Number built** 410.

Sir Nigel Gresley died in office during April 1941 and was succeeded as Chief Mechanical Engineer of the LNER by Edward Thompson. Gresley's big engine policy had left that railway well provided for in terms of power on the main routes, but the secondary routes were sorely tried under the prevailing wartime conditions. Upon assuming control, Thompson cancelled an order for ten 'V4' 2-6-2 locomotives, the final Gresley design intended to operate secondary routes, considering them to be too complicated. He was impressed by the success of the GWR 'Halls' and LMS 'Black Fives' and was determined to produce an LNER version. That his decision was correct is undeniable, as normal operations and maintenance suffered due to the war.

The Thompson 'B1' Class would be as simple and easy to maintain as possible. Austerity would be the keyword in design, but as many standard LNER components as practicable would be incorporated. The appearance of the first 'B1', named *Springbok*, was a pleasant surprise to many who feared that the austerity bug caught by Bulleid with his 'Q1' might have infected Thompson. In fact, the new locomotive looked very attractive and typically LNER with its parallel boiler and round-topped firebox. A two-cylinder arrangement had been chosen rather than the three cylinders which had been Gresley's practice, and obviously maintenance and preparation would be easier with no inside cylinder. To reduce the 'hammer blow' effect consequent upon the two-cylinder design, only 30% of the reciprocating masses were balanced. This had the effect of reducing the total loading on bridges compared with a three-cylinder 'B17' of Gresley design with an equivalent tractive effort. However, the price paid for using only two cylinders was rough riding due to the imbalance of the reciprocating masses.

Electric lighting was applied to a number of the class but poor maintenance soon resulted in defects, an example of a good idea rendered bad by post-war labour shortages. Another useful

Above *'B1' Class No 61237 at York on 7 September 1965* (D. K. Jones Collection).

Below right *'B1' Class No 61269 at Gorton, 17 March 1961.*

Below *Twenty-one years later, preserved 'B1' No 1306, now named* Mayflower, *is seen at Loughborough, 7 September 1986.*

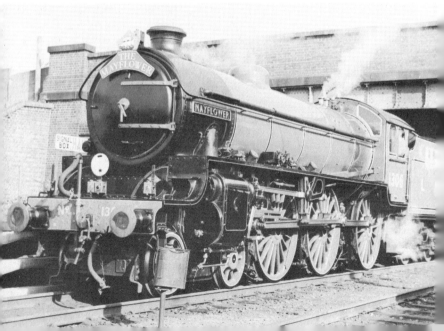

feature fitted to later members of the class was the self-cleaning smokebox as fitted to the 'Black Fives'; it was intended to reduce shed maintenance but it impaired steaming and showered hot ash over the locomotive and lineside. Damage to clothing and eye injuries were frequently the result.

On the positive side, the 'B1' cabs were fitted with bucket seats, more comfortable than the stools on similar engines. They were also strong and reliable, being capable of hauling heavy loads up the most severe of gradients. The locomotive exchange trials of 1948 pitted the 'B1's against the 'Black Fives' and GWR 'Halls', and the results showed that Thompson had produced a very creditable design. Considering that it originated during the Second World War and had to contend with existing production and operating circumstances, it was an excellent piece of engineering.

Only ten 'B1s' were actually constructed in LNER workshops before 1945, but when post-war replacement of older stock demanded accelerated production it was necessary to deal with contractors. The North British Locomotive Co and the Vulcan Foundry built large numbers, with those from Vulcan being considered the best 'B1s'.

Two have escaped the cutter's torch and both are at the great Central Railway, Loughborough, although No 61264 required extensive work to rectify its boiler defects before steaming could be attempted.

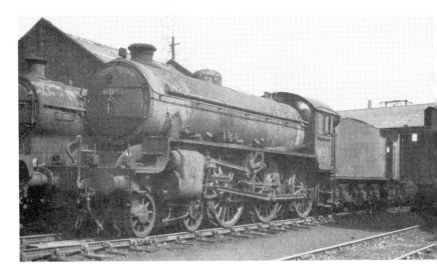

LMS REBUILT 'ROYAL SCOT' CLASS 4-6-0

Designer North British Locomotive Co/W. A. Stanier. **Introduced** 1943 (from the 1927 original). **Power classification** 7P. **Driving wheel diameter** 6 ft 9 in. **Leading wheel diameter** 3 ft 3.5 in. **Cylinders** Three, 18 in diameter × 26 in stroke. **Valve gear** Walschaerts. **Boiler**—*Pressure* 250 psi. *Grate area* 31.2 sq ft. **Heating surfaces**—*Tubes* 1,667 sq ft. *Firebox* 195 sq ft. *Superheater* 357 sq ft. **Tractive effort** 33,150 lbs. **Weight** Loco 83 tons, tender 54 tons 13 cwt. **Number built** 70.

The Midland Railway's dominance of the LMS in the years following the grouping had a disastrous influence on the passenger locomotive stock. Relatively small locomotives abounded, with the 'Compounds' being favoured even when double-heading became the only way of hauling the increasingly heavy trains. Matters came to a head when the GWR's *Launceston Castle* put LMS engines to shame on their home ground. The directors ordered Henry Fowler to rectify matters immediately, but it was not possible. Outside help came to the rescue and, in February 1927, the North British Locomotive Co was given an order for fifty express passenger locomotives using drawings of the Southern's 'Lord Nelson' Class as a basis; the detail design work was undertaken by the North British in consultation with Derby. The first 'Royal Scot' was delivered on July 14 1927 and the final member of the batch on November 15, a design and manufacturing achievement which must be something of a record. A further twenty 'Scots' were produced at Derby before the end of the decade.

The 'Scots' proved to be a success; at least they were better than any other LMS passenger locomotive although in later years they looked very jaded alongside Stanier's machines. In 1943, a programme of rebuilding commenced but it was not completed until 1955 to allow direct comparison between originals and rebuilds; the popular conclusion was that the reconstruction made good locomotives into great ones.

A taper boiler replaced the very square looking unit of the original design, but other less eye-catching features also changed. In fact, they were almost new machines with only the cab, wheels and some smaller items being retained from the originals. The tender had already been replaced by one of Stanier design some years earlier.

The rebuilding was based on the 1935 Stanier reconstruction of Fowler's experimental engine *Fury*. A tragic boiler explosion resulted in premature curtailment of the experiments, and the locomotive lay idle for many years until Stanier made use of it to try out his own modification plans. Although not strictly a 'Royal Scot', it was added to the class for allocation and work purposes, essentially making the class total 71.

The relatively light axle loading of 20 tons 9 cwt gave the class access to many routes not available to Stanier's 'Pacifics', and it was here they proved their worth. Even on main-line express trains from Euston they could show a turn of speed and were admirable substitutes for their larger 'Pacific' sisters. The Stanier taper boiler gave them a high reputation for steaming, but drifting smoke was an occasional problem. The fitting of smoke deflectors cured this, although many thought they detracted from their appearance. Under certain conditions the riding of the 'Scots' could be rough, but it was uncomfortable rather than dangerous.

As express passenger locomotives, the original 'Royal Scots' were good, but following reconstruction, as the rebuilding was officially termed, they became something extra. Few locomotive classes have had two existences, but the second was worth waiting for.

Two rebuilt 'Scots' are preserved, No 6100 *Royal Scot* is in the capable hands of Alan Bloom at Bressingham and No 6115 *Scots Guardsman* is at Dinting.

Rebuilt 'Royal Scot' No 46122 Royal Ulster Rifleman *passing Camden on 7 June 1952* (D. K. Jones Collection).

WAR DEPARTMENT 'AUSTERITY' 0-6-0 SADDLE TANK

Designer Hunslet Engine Co. **Introduced** 1943. **Power classification** 4F. **Driving wheel diameter** 4 ft 3 in. **Cylinders** Two (inside) 18 in diameter × 26 in stroke. **Valve gear** Stephenson. **Boiler**—*Pressure* 170 psi. *Grate area* 16.75 sq ft. **Heating surfaces**—*Tubes* 873 sq ft. *Firebox* 87 sq ft. **Tractive effort** 23,870 lbs. **Weight** 48 tons 3 cwt. **Number built** 484.

During the late 1930s, a heavy 0-6-0 saddle tank, the '50550' Class, was developed by the Hunslet Engine Company to a special order from Stewart & Lloyds. These locomotives were intended for use at that company's Corby steel works but the Second World War interrupted delivery. As the war progressed, it became obvious that large numbers of powerful shunting engines would be required for use in this country and in Europe following an invasion. Initially the LMS 'Jinty' was selected, but Edgar Alcock, Chairman of Hunslets, convinced the War Department that a locomotive based upon the '50550' would be more suitable. Subsequently Hunslets received an order for a large number of such engines and so the basic 'Austerity' 0-6-0ST was born.

Robert Riddles was given the task of modifying the existing design for war use, but these changes mainly amounted to material specifications for cheapness and speed of manufacture. Hunslet could not manage the entire order themselves and other locomotive builders including Robert Stephenson & Hawthorn, Bagnall, Hudswell Clarke and Vulcan turned out large numbers. Throughout the war years, these powerful little machines did all that was asked of them and they more than lived up to their expectations. Being of relatively simple design and construction, servicing presented few problems making them extremely popular with their operators. They were, in effect, the ideal locomotive for the multifarious duties imposed by the military, and their reliability could not be faulted when compared with similar locomotives.

Following the end of hostilities, surplus locomotives were sold by the army, with the LNER taking 75 for shunting duties in yards throughout its system but mainly in the North East. These were reclassified as 'J94' and many remained in service until the end of steam on BR Eastern Region. Other surplus engines went to industrial operators, particularly the National Coal Board who found them to be so suitable for the work that they ordered more from

0-6-0 'Austerity' posing as BR locomotive No 68009, although this engine never actually operated for BR.

Hunslets. The last was not completed until 1964, four years after BR had finished steam locomotive construction. A large number remained with the army, operating in this country and abroad, while some were sold for industrial service in Holland and two are now preserved in that country. Withdrawals were gradual from both military and industrial use, as these 'Austerities' were so useful and economical.

As industrial diesel locomotives became more popular and British industry began to suffer a general run-down, the final demise of the class became inevitable, but they did not cease commercial service until the 1980s. Because of the 'Austerity's survival in industrial use, a large number of examples have been preserved, many being taken directly from service with the NCB.

These little engines could pull the heaviest of loads over the roughest of colliery and other tracks whilst requiring only minimal maintenance. In short, they fulfilled their design requirements and then gave more, making them indeed 'a work of recognized excellence'.

WAR DEPARTMENT 'AUSTERITY' 2-10-0

Designer R. A. Riddles. **Introduced** 1943. **Power classification** 9F. **Driving wheel diameter** 4 ft 8.5 in. **Leading wheel diameter** 3 ft 2 in. **Cylinders** Two (outside) 19 in diameter × 28 in stroke. **Valve gear** Walschaerts. **Boiler**—*Pressure* 225 psi. *Grate area* 40 sq ft. **Heating surfaces**—*Tubes* 1,759 sq ft. *Firebox* 192 sq ft. *Superheater* 423 sq ft. **Tractive effort** 34,210 lbs. **Weight** Loco 78 tons 6 cwt, tender 55 tons 10 cwt. **Number built** 150.

With the outbreak of the Second World War, R. A. Riddles was appointed director of transport equipment at the Ministry of Supply, one of his responsibilities being procurement of locomotives. Immediate needs were met by the GWR 'Dean Goods' but a more powerful locomotive was required for service in Europe and the other theatres of war. It was essential that a fairly standard type of engine should be chosen, and the Stanier '8F' 2-8-0 certainly fitted the bill, over 200 being subsequently constructed for WD service. Allied bombing was partly aimed at immobilizing the railways in German hands and many locomotives on the European mainland were consequently destroyed. Obviously an extensive fleet would have to be constructed to support an invasion force.

Riddles decided to construct an entirely new engine rather than build further '8Fs', for two very good reasons. Firstly, the Stanier design required a considerable number of man hours to construct,

Riddles 'Austerity' 2-10-0 No 600 Gordon *operating on the Severn Valley Railway.*

and secondly it made use of valuable materials which were no longer readily available. The Riddles-designed locomotive would be a 2-8-0 but of simplified construction using fabrication as far as possible rather than castings. Access for maintenance was made as simple as possible and although the locomotive had a very basic look it was functional and effective. Fabrication and the elimination of all but essential components allowed an axle loading of only 15 tons 12 cwt. The Vulcan Foundry and the North British Locomotive Co supplied some 935 locomotives of this design between them during the 1943-45 period.

Although the 2-8-0s were ideal for most purposes, it was evident that an equally powerful engine with a lower axle loading would be essential for use on track of doubtful quality such as might be expected following the invasion, so Riddles set to work modifying his earlier design. A 2-10-0 arrangement gave a lower axle loading, but, for the locomotive to negotiate curves of 4.5 chains radius without derailment, the centre coupled wheels had to be flangeless whilst those on either side of centre had flanges of reduced thickness. The longer total wheelbase allowed for a longer boiler which had a wider firebox with a rocking grate. As with the 2-8-0s, the steel firebox design allowed for easy conversion to oil firing should that be necessary.

Because the cylinder dimensions, coupled wheel diameter and boiler pressure remained the same as the eight-coupled engines, the tractive effort was also identical. It was, however, the low axle loading of only 13 tons 10 cwt which gave the larger locomotive its degree of superiority. North British built all 150 members of the class; some of them never in fact left Britain whilst others found service in Europe and the Middle East. Following demobilization, 25 2-10-0s were taken into BR stock and two stayed in WD ownership. A number remained active in the Middle East and in Greece until the late 1970s.

One of the WD pair, WD73651, now named *Gordon*, resides on the Severn Valley Railway whilst the other, WD73755, named *Longmoor*, occupies a museum place in Holland. Two members of the class were returned from Greece during 1984 and are now to be found on the Mid-Hants and North Yorkshire Moors Railways.

These locomotives may not have had the glamour of the Stanier or Gresley 'Pacifics' but they were inexpensive to construct and easy to maintain. They also fulfilled their design requirement and it was from them that the final British steam design, the standard '9F', originated.

LNER 'A2' CLASS 'PACIFIC' 4-6-2

Designer E. Thompson/A. H. Peppercorn. **Introduced** 1943/1947. **Power classification** 8P7F. **Driving wheel diameter** 6 ft 2 in. **Leading wheel diameter** 3 ft 2 in. **Trailing wheel diameter** 3 ft 8 in. **Cylinders** Three, 19 in diameter × 26 in stroke. **Valve gear** Walschaerts. **Boiler**—*Pressure* 250 psi. *Grate area* 50 sq ft. **Heating surfaces**—*Tubes* 1,217 sq ft. *Firebox* 245 sq ft. *Superheater* 680 sq ft. **Tractive effort** 40,430 lbs. **Weight** Loco 101 tons, tender 60 tons 7 cwt. **Number built** 15 (The figures given are for the final Peppercorn version).

On the death of Sir Nigel Gresley in 1941, Edward Thompson became Chief Mechanical Engineer of the LNER and a new phase of locomotive development began. Many people have claimed that he set out to deliberately overturn his predecessor's works and there may be some truth in that. However, war conditions were different from those encountered by Gresley and it was Thompson's task to deal with the circumstances of the day. Conjugated valve gear suffered if it was not maintained in first class order and wartime did not allow for effective maintenance. It was thus dispensed with, and for future three-cylinder engines the inside cylinder would have its own set of Walschaerts gear. Thompson rebuilt as 'Pacifics' Gresley's 'P2' Class 2-8-2 locomotives of 1934 and these became the 'A2/2' Class.

The 'A2/2s' were three-cylinder engines in which the inside cylinder drove the forward coupled axle and the outside cylinders the centre axle. Thompson had a preference for connecting rods of equal length and to achieve this the inside cylinder had to be moved as far forward as possible whilst the outside cylinders were moved towards the rear. The connecting rods were actually retained from the 'P2s' in order to reduce the number of new parts required. Also, the large diameter cylinders had to be positioned as far back as possible to give clearance for the bogie wheels, and this produced a rather strange appearance—they looked wrong, as if the bogie had been something of an afterthought. Four 'V2s' then under construction were similarly 'rebuilt' and emerged as 'A2/1s', and further locomotives, this time 'A2/3s', were constructed in 1946.

Thompson was unpopular because of the way he changed the old system, and many enthusiasts were pleased when he retired in July 1946 to be replaced by A. H. Peppercorn. Although nationalization was planned, it was 18 months away and several

Above *Preserved 'A2' 'Pacific'* Blue Peter *displaying apple green livery and LNER No 532, neither of which it ever carried in service.*

Below *An 'A2' in BR days –* No 60533 Happy Knight *at Doncaster on 18 July 1961* (D. K. Jones Collection).

'Pacifics' were still on order. The new CME decided that changes would be made, especially with regard to the equal-length connecting rods and the positioning of the outside cylinders. That certainly removed a number of defects, particularly with regard to the flexing of the frames and steam-pipe connections to the outside cylinders. One Gresley feature was not restored, however—there would be no return to the conjugated valve drive for the inside cylinder. The Peppercorn 'A2' 'Pacifics' certainly looked more attractive than their predecessors and mechanically they were excellent with a record of high availability. In 1949, five of the later engines were given Kylchap double blastpipes and that produced a marked improvement in steaming ability as well as fuel economy.

In retrospect, the Peppercorn 'A2s' had the edge over their earlier sisters but both groups were fast and free running. The Thompson engines were, however, bad riders because of the frame flexibility and they had an unenviable record for mechanical faults. Using the doubtful tractive effort criteria, the 'A2s' were amongst the most powerful locomotives to operate in Britain and the surviving example, No 60532 *Blue Peter* is the most powerful locomotive in preservation. That alone merits their inclusion in this book.

Considering the fact that he had so little time in which to make the necessary changes, Peppercorn certainly motivated his drawing office team to turn a rather inferior Thompson design into one of which Gresley himself would have been proud. Powerful, reliable, economical and possibly even beautiful, the 'A2s' were appropriate engines with which to draw to a close the history of LNER locomotive design.

'A2' Class 'Pacific' No 60528 Tudor Minstrel *at Haymarket, 11 November 1961.*

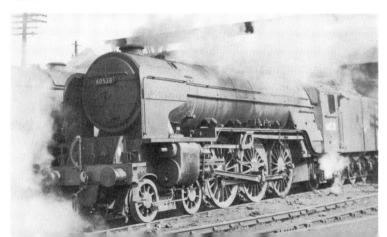

SOUTHERN RAILWAY 'WEST COUNTRY' AND 'BATTLE OF BRITAIN' CLASS 'LIGHT PACIFIC' 4-6-2

Designer O. V. S. Bulleid. **Introduced** 1945. **Power classification** Originally 6P, Later 7P5F. **Driving wheel diameter** 6 ft 2 in. **Leading wheel diameter** 3 ft 1 in. **Trailing wheel diameter** 3 ft 1 in. **Cylinders** Three, 16.375 in diameter × 24 in stroke. **Valve gear** Bulleid patent chain driven (Walschaerts). **Boiler**—*Pressure* 280 psi (250 psi). *Grate area* 38.25 sq ft. **Heating surfaces**—*Tubes* 1,869 sq ft. *Firebox* 253 sq ft. *Superheater* 545 sq ft (488 sq ft). **Tractive effort** 31,000 lbs (27,720 lbs). **Weight** Loco 86 tons 0 cwt (91 tons 1 cwt), tender 42 tons 12 cwt. **Number built** 110. (Figures in brackets are for the BR rebuilt version where different.)

The 'West Country' and 'Battle of Britain' Class 'light Pacifics' were identical in every respect, but the Southern Railway considered there to be a public relations advantage in naming a group of its latest locomotives after people and places associated with the aerial conflict to defend Britain. Before the Second World War, Bulleid had convinced the railway's directors that an improved stock of steam locomotives was required, despite the existing policy of electrification. He obtained permission to construct ten main-line steam locomotives and these eventually became the first of his 'Merchant Navy' Class 'Pacifics'.

They were revolutionary in style and operation, particularly with respect to the valve gear; this unorthodox chain driven arrangement contained within its own oil-bath was designed to ease maintenance. An air-smoothed casing gave a streamlined effect, but the desire was not to reduce wind resistance but to allow cleaning of the locomotives in a carriage washing plant. At 280 psi, the boiler pressure was higher than that applied to any other production class of British locomotive. Wheels of the patent Bulleid-Firth-Brown type were used in order to limit the total locomotive weight. In concept, the design was good, but the outbreak of war did not allow sufficient time for testing and modification.

The 'Merchant Navy' Class locomotives were barred from many of the secondary routes because of weight restrictions, so Bulleid was able to convince those in authority that a lighter version should

Unrebuilt Bulleid 'Light Pacific' No 34092 City of Wells *at Appleby with an enthusiasts' special.*

be produced. This was sanctioned despite the teething troubles of the larger class and the wartime embargo on construction of any but freight locomotives—by classing the machines as 'mixed traffic' that difficulty was circumvented. The first 'light Pacific' did not leave Brighton works until June 1945 and looked like a scaled-down version of the 'Merchant Navy' 'Pacifics'. In fact they were just that.

The cylinder diameter had been reduced by 1.625 in but the stroke remained the same. The boiler dimensions were smaller to allow for a reduction in locomotive length and weight, but in other respects the same features were employed, including 6 ft 2 in diameter B-F-B patent wheels. Following smoke drift problems with the larger 'Pacifics', modifications to the boiler casing front end had been made and that new arrangement was used. These new locomotives actually had a tractive effort below that of the older 'Lord Nelsons' but the high pressure boiler allowed for a superior sustained pulling power, and the low axle loading gave an extensive route availability.

There were still problems with the 'light Pacifics', but not nearly so many as with their larger sisters, and these were mainly but not entirely due to the valve gear. The boiler, however, was excellent and it is doubtful if a better steam generator has ever been produced for operation on a British locomotive. Some crews loved the machines whilst others loathed them, but some of the problems were due to an inability to master the driving techniques needed. To misquote a well-known nursery rhyme, 'When they were good they were very, very good, but when they were bad they were horrid'. During the 1948 interchange trials, the 'West Country' representatives put in some sterling performances but at the price of high fuel consumption. At other times, especially when in a run-down condition, they performed abysmally. It might be said, however, that the 'light Pacifics' were less bad than the 'Merchant Navy' Class 'Pacifics'.

The non-standard nature of the locomotives together with their heavy oil and coal consumption conspired against them and a decision was taken to reconstruct them in a more conventional form. The valve gear was replaced by a set of Walschaerts for each cylinder and the air-smoothed casing removed to restore the usual locomotive shape. A new inside cylinder replaced the original, the smokebox, superheater header and steam-pipes were changed, a hand-operated screw reverser was fitted instead of the steam reverser and steam operation of the firehole door was removed. Other small modifications were also made.

All of the 'Merchant Navy' 'Pacifics' were rebuilt, but only 60 of the 'light Pacifics' were so treated before a halt was called to the process due to the impending abandonment of steam. Debates still take place regarding the merits of these locomotives in their original and rebuilt forms, but there can be little doubt that the 'light Pacifics' were better than the heavier variety. Fortunately, examples of both have been preserved to allow latter-day enthusiasts to compare shapes if not performances.

Rebuilt Bulleid 'Light Pacific' No 34016 Bodmin, *now preserved on the Mid-Hants Railway.*

LMS '4MT' CLASS 2-6-0

Designer H. G. Ivatt. **Introduced** 1947. **Power classification** 4F, later 4P/4F, 4MT. **Driving wheel diameter** 5 ft 3 in. **Leading wheel diameter** 3 ft 0 in. **Cylinders** Two (outside) 17.5 in diameter × 26 in stroke. **Valve gear** Walschaerts. **Boiler**—*Pressure* 225 psi. *Grate area* 23 sq ft. **Heating surfaces**—*Tubes* 1,090 sq ft. *Firebox* 131 sq ft. *Superheater* 246 sq ft. **Tractive effort** 24,555 lbs. **Weight** Loco 59 tons 2 cwt, tender 40 tons 6 cwt. **Number built** 162.

Charles Fairburn died suddenly in October 1945, and in January 1946 H. G. Ivatt, son of the famous GNR Locomotive Engineer, was appointed to succeed him as Chief Mechanical Engineer of the LMS. The approaching nationalization of the railways meant that he would be the last person to hold the post. Ivatt did not attempt to impose his own ideas on the locomotive fleet or change the policies of his predecessors, as some engineers on other railways had done, but he did make use of improved methods of construction where possible. He also introduced ideas aimed at simplifying maintenance, particularly as changed circumstances following the war had reduced the number of people willing to undertake the dirty shed jobs.

Ivatt Class '4' 'Mogul' No 43098 at Normanton, 2 September 1966 (D. K. Jones Collection).

'The Flying Pig', preserved Ivatt 'Mogul' No 43106, at Bridgnorth on the Severn Valley Railway.

Most famous of his three steam designs was the Class '4' 'Mogul', originally designated as a goods engine. It was intended as a replacement for the numerous 0-6-0 goods engines which abounded throughout the system but ultimately turned out to be much more useful. The influence of American practice was evident in the design, especially that of the 'S160s' which had been brought over as part of the American buildup to the invasion of Europe. The high running-plate was attached to the boiler flanks thus leaving the wheels and frames completely exposed. This gave the locomotive a rather ugly appearance in some eyes but it improved accessibility for maintenance, and Ivatt was more concerned about his designs being functional than good-looking.

The smokebox was self-cleaning and the ashpan self-emptying in order to reduce the work required on-shed. The footplate crew was not forgotten, as the tender was provided with a cab to afford protection when running tender first with an inset coal bunker allowing for a clear view, and, with exposed running gear, the routine oiling and preparation checks were simplified. Initially, a large double chimney was provided, accentuating the rather austere, unorthodox design of the engine, but in practice it was found to be ineffective due to incorrect dimensioning, and a single chimney soon followed.

The boiler itself was an excellent steamer and following tests at Swindon in 1951 an improved draughting arrangement increased its effectiveness even more. The LMS batches were constructed at Horwich, but after nationalization further members of the class were built at Doncaster and Darlington for use on the Eastern and North Eastern Regions of British Railways. Few people had any complaints about the design once the initial draughting problems had been corrected.

The advent of the BR standard classes only served to illustrate that Ivatt's design was good and required no engineering improvement. The standard Class '4' 2-6-0 was basically the same locomotive modified to include certain standard details and made visually more attractive. This last feature was unnecessary, but it gave some uniformity of appearance to the standard classes.

Although originally designed as freight engines, these Ivatt machines soon showed themselves to be effective on all types of trains and their designation was changed to that of mixed traffic. High-speed running with passenger and special trains earned them the nickname 'Flying Pig' which has stuck to the one surviving example, No 43106, now on the Severn Valley Railway.

Locomotives must not only be effective in traffic but they must be designed to suit construction and maintenance facilities available at the time. With his Class '4' locomotive, Ivatt did that admirably and pointed the way to the standard designs of later years.

'4MT' Class No 43151 at Crewe South, 1 July 1966.

BR '9F' CLASS 2-10-0

Designer R. A. Riddles. **Introduced** 1954. **Power classification** 9F. **Driving wheel diameter** 5 ft 0 in. **Leading wheel diameter** 3 ft 0 in. **Cylinders** Two (outside) 20 in diameter × 28 in stroke. **Valve gear** Walschaerts. **Boiler**—*Pressure* 250 psi. *Grate area* 40.2 sq ft. **Heating surfaces**—*Tubes* 1,835 sq ft. *Firebox* 179 sq ft. *Superheater* 535 sq ft. **Tractive effort** 39,670 lbs. **Weight** Loco 86 tons 14 cwt, tender 52 tons 10 cwt. **Number built** 251.

The locomotive standardization scheme following nationalization of the railways called for a heavy freight engine. Initially, a 2-8-2 wheel arrangement was proposed and some design work was carried out, but during the late 1940s a heavy freight engine was not a priority, with many war-built 2-8-0s available, so construction did not actually take place. Riddles also appears to have had second thoughts as to the suitability of an eight-coupled locomotive in view of the good performances of his 2-10-0 locomotives designed for the War Department. Eventually, a 2-10-0 arrangement was settled on and the design brief given to the drawing office at Brighton. As a post-nationalization policy, each of the major railway company drawing offices was to be given overall responsibility for the design of certain standard classes.

Few parts from the other standard classes could be used and the '9Fs' essentially became an isolated group, thus destroying one of the intended advantages of the standardization policy. Within the class itself there were a number of different design features, the most dramatic being the use of a Crosti boiler on some engines. The Crosti principle was adopted as a means of improving efficiency by preheating the boiler feed water. Instead of the gases escaping directly up the chimney they were directed through the tubes of a heat exchanger. This drum unit was placed below the boiler and the feed water circulated around the tubes through which the flue gases passed. The gases then exhausted on the right-hand side of the locomotive just in front of the firebox. To accommodate the Crosti arrangement, the boiler itself had to be smaller than normal in order to fit within the loading gauge. As an idea to save coal during a period of fuel crisis it was admirable, but it did not work well in practice. It was not as economical as expected, maintenance costs were higher and major corrosion problems developed in the smokebox, so after four years the experiment was discontinued and the locomotives modified.

Some members of the class were provided with a double

The final steam locomotive built for BR, No 92220 Evening Star, *at Craven Arms in 1983.*

chimney which proved useful in enhancing steaming under some working conditions and it improved the appearance of an already good-looking locomotive. No 92250 was fitted with a Geisel oblong ejector in 1959 to test out the merits of that device but, as with the Crosti boiler, results proved disappointing. Other engines were fitted with mechanical stokers in order to increase the firing rate above that which could be expected from a single fireman and thus improve the evaporation rate. Results showed that a 10% increase in evaporation was possible, but with an accompanying rise in coal consumption. Although specially crushed coal was required, the system did have possibilities for very heavy trains but, again, the experiment was discontinued. This time the reason lay not with any system imperfections but because of dieselization and the impending demise of steam.

All things considered, the '9Fs' were very good, hard-working and efficient locomotives, arguably one of the best designs ever to operate in Britain. Their biggest drawback was that they came too late, or rather steam finished too soon. Some '9Fs' went to the scrapyard after barely six years' service, an economic folly of massive proportions. Fortunately, David Shepherd purchased No 92203, now called *Black Prince*, from BR in 1967 whilst No 92220 *Evening Star*, the last steam locomotive constructed for British Railways, was set aside for the National Collection. In recent years, other '9Fs' have been purchased from Barry scrapyard for ultimate restoration.

BR '8P' CLASS 4-6-2 *DUKE OF GLOUCESTER*

Designer R. A. Riddles. **Introduced** 1954. **Power classification** 8P. **Driving wheel diameter** 6 ft 2 in. **Leading wheel diameter** 3 ft 0 in. **Trailing wheel diameter** 3 ft 3.5 in. **Cylinders** Three 18 in diameter × 28 in stroke. **Valve gear** British Caprotti. **Boiler**—*Pressure* 250 psi. *Grate area* 48.6 sq ft. **Heating surfaces**—*Tubes* 2,264 sq ft. *Firebox* 226 sq ft. *Superheater* 677 sq ft. **Tractive effort** 39,080 lbs. **Weight** Loco 101 tons 5 cwt, tender 53 tons 14 cwt. **Number built** 1.

The original plans for the BR standard locomotives made provision for a large express passenger 'Pacific' but no immediate need for such a machine existed. The Harrow crash in which former LMS 'Pacific' No 46202 was destroyed left a gap in the 8P locomotive fleet and Riddles sought a replacement. Permission was granted for construction of a 'Pacific' along the lines of that previously envisaged, thus allowing valuable running experience to be gained before full scale production. No 71000 *Duke of Gloucester* was completed at Crewe in 1954 and set to work on the West Coast main line.

'The Duke' differed from the standard Class '6' and Class '7'

Duke of Gloucester *minus outside cylinders, chimney, rods and other parts at Cashmore's scrapyard in 1967* (D. K. Jones Collection).

Fully restored, No 71000 Duke of Gloucester *in steam on the Great Central Railway, 7 September 1986.*

'Pacifics' in that it had three cylinders instead of two and Caprotti valve gear rather than the Walschaerts type. The inside cylinder drove the leading coupled axle and rotary cam poppet valves avoided the problems normally associated with valve gear drive for the centre cylinder of a three-cylinder engine. The boiler was similar to that of the 'Britannia' Class 'Pacifics' but it had a larger grate together with a double chimney and blastpipe. The wheels had roller bearings, and in other details also No 71000 followed 'Britannia' practice. The tender was unique, having a coal capacity of 10 tons and a water capacity of 4,325 gallons; it was also provided with a steam-operated coal pusher.

In service, *Duke of Gloucester* failed to impress, being heavy on coal and lacking the punch of the Stanier 'Pacifics'. The decision to cease steam operations curtailed any further experimentation or redesign which could have improved matters and allowed the engine to live up to its design expectation. Following withdrawal in 1962, only the outside cylinders and Caprotti valve gear were preserved for exhibition, the remainder of the machine being sold to Woodhams of Barry for scrap. No 71000 actually turned up at Cashmore's scrapyard in Newport and stripping commenced. Fortunately, an enthusiast noticed the label indicating the purchasers as being Woodhams and the derelict locomotive was redirected. But for that chance encounter, 'The Duke' would have been cut up.

A small group of dedicated enthusiasts decided that the engine should be preserved and after considerable effort raised the purchase price of £4,950. A move to the Great Central Railway at Loughborough was only the start, for vitally important parts of the engine were missing. It was a technical challenge way beyond that undertaken by any other preservation group, because new cylinders, valve gear and rods, amongst the more usual items, would be required. No other group had previously had to start from scratch with such major items and at first the project was treated as a joke. However, perseverance, enthusiasm and dedication saw the project through, but it was a long hard struggle which set the group above any other in locomotive preservation. Contributions from the enthusiast fraternity were invaluable, as was the help provided by some sections of industry, but without the little band of people who fought on against the odds there would now be no *Duke of Gloucester* to admire.

The unique nature of the construction and performance potential of No 71000 alone might merit inclusion in a list of classic locomotives but that is not the reason here. *Duke of Gloucester* has its section in this book because of the classic way in which it has been restored. Nobody could argue that the efforts of the preservation team have not been worthwhile, or underestimate the enormity of the task they undertook and completed. They may even have solved one of the mysteries of 'The Duke's' performance. With no drawing available, a new ashpan was constructed using the old one as a pattern, but when the drawings became available it was obvious that the actual damper door area was smaller than originally intended. At high outputs, that restriction in area would be significant and it is to be hoped that one day this classic example of locomotive preservation will be able to show its true capabilities.

BR DIESEL-ELECTRIC CLASS '40'

Designer English Electric. **Introduced** 1958. **Driven wheel diameter** 3 ft 9 in. **Pony wheel diameter** 3 ft 0 in. **Axle arrangement** 1Co-Co1. **Engines** EE16 SVT Mk11. **Power** 1,480 kW (2,000 hp) at 850 rpm. **Type of drive** Electric, six E.E.526/5D axle-hung motors. **Tractive effort** 52,000 lbs (maximum), 30,900 lbs (continuous at 18.8 mph). **Maximum speed** 90 mph. **Full weight** 133 tons. **Number built** 200.

The initial railway modernization plan included a dieselization scheme with main-line locomotives in four broad categories of which Type '4' covered diesels of 2,000 hp and above. English Electric's contribution to the Pilot Scheme in the Type '4' category consisted of ten 'D200' Class, later Class '40', locomotives. A turbo-charged 16-cylinder English Electric SVT engine was used in order to meet the power requirement and a Clayton RO2500 boiler provided steam for train heating. The high total weight of 133 tons dictated a 1Co-Co1 wheel arrangement to keep the axle loading to 20 tons. Gangway doors were fitted in the nose at each end of the locomotive in order to provide a connection, through suitable bellows, when double-heading. This feature was retained for some of the early production engines but was rarely used and later locomotives were not so equipped.

The ten Pilot Scheme Class '40s' performed well on trials and large-scale production was sanctioned. So satisfactory had the original ten been that few changes were required in the production version. In May 1959, these machines entered service on the London Midland main line from Euston to the north. So effective and reliable were they that consideration was given to upgrading them to 2,400 hp but nothing came of the proposal. Early problems with traction motor flash-over caused concern, but this was overcome by replacing the original six-pole motors with four-pole motors. Batches were allocated to the Eastern, North Eastern and Scottish regions, all giving a good account of themselves.

The Class '40s' were at the lower end of the Type '4' power range but they proved to be one of the best designs in that category, being reliable and having a high availability. That reliability remained throughout the class's service life, which is more than can be said for many of the diesel classes ordered in haste under the blanket of modernization. Electrification of the West Coast main line and production of more powerful diesels resulted in the '40s' being directed to less glamorous secondary

No 1 end of Class '40' No 40 093 showing the front communicating doors and headcode discs.

passenger and freight work. They were, however, equally at home on that type of service and would often be preferred because of that high reliability. Being one of the original diesel classes, maintenance facilities in the early years frequently had to be shared with steam locomotives. Rugged construction and sound design allowed the '40s' to shrug off such conditions whilst other diesels suffered.

The characteristic whistle produced by the large engine-driven fan soon resulted in the nickname 'whistler' being applied to the entire class. That unmistakable sound made the presence of an unseen '40' obvious to any enthusiast. By the early 1980s, withdrawals were being made at an accelerating rate with elimination of the class from BR metals by the end of 1984. However, the Class '40s' were to prove more durable than that, and as their numbers diminished the adulation grew, with enthusiasts making pilgrimages to many parts of the country in order to see or ride behind those which remained. Even BR seemed reluctant to finally rid itself of the most successful Pilot Scheme diesel, and a number were taken into departmental service following nominal withdrawal. These Class '97s' were, however, still used on service trains whenever necessity dictated, a fitting tribute to their reliability.

A number of examples of the class have been preserved so that the younger generation may see and hear a pioneering diesel which was a worthy successor to the steam age.

BR 'DELTIC' CLASS '55'

Designer English Electric. **Introduced** 1961. **Wheel diameter** 3 ft 9 in.
Axle arrangement Co-Co. **Engines** Two, Napier 'Deltic' 18-25 series.
Power 2 × 1,230 kW (1,650 hp) at 1,500 rpm. **Type of drive** Electric, six
E.E.538 axle-hung motors. **Tractive effort** 50,000 lbs (maximum), 30,500
lbs (continuous at 32.5 mph). **Maximum speed** 100 mph **Full weight** 100
tons. **Number built** 22.

The Napier 'Deltic' engine derives its name from the fact that its
cylinders are in the triangular form of an inverted Greek letter delta.
There are 18 cylinders in each engine arranged in six triangular
groups of three; two opposed pistons work in each cylinder driving
crankshafts in each angle of the 'triangle' on a two-stroke cycle.
The three crankshafts are then geared to a single drive shaft. The
arrangement was originally developed as a high-power lightweight
engine for driving naval patrol boats, but its potential as a rail
traction unit soon became evident. As the British Railways
modernization plan began to develop, English Electric decided to
promote its own locomotive based upon the 'Deltic' engine. They
constructed a prototype costing £250,000 at their own expense
and arranged for extensive trials on BR lines. It was on the East
Coast main line that the full potential was realized, and
investigations showed that a small fleet of 22 'Deltics' could be
used to replace 55 steam locomotives and allow for a speeding up
of schedules. The BR board was convinced and in 1958 approved
the purchase of 22 locomotives at a cost of £200,000 each.

Deliveries commenced in May 1961, and services on the former
LNER routes from King's Cross to the north were gradually
accelerated. Though the travelling public were appreciative of the
new services, railway enthusiasts did not take kindly to the
interlopers. They had, after all, doomed many well-loved 'Pacifics'
to the scrapyard. However, as steam finally disappeared from
British main lines, the 'Deltics' became more appreciated and
eventually something approaching a 'cult following' developed.

Three trains, the 'Elizabethan', 'Talisman' and 'Flying Scotsman', all offering daily six-hour schedules between London and
Edinburgh, were the highlights of the new schedules. The
'Tees-Tyne Pullman' offered what was then the highest ever
average point-to-point speed of any train in Britain, 35 minutes
between Darlington and York at 75.6 mph. Over the ensuing years,
the track was improved allowing the 'Deltics' to operate closer to
their full potential for longer periods. Times for the Scottish

services were reduced whilst the timing of the 'Tees-Tyne Pullman' between Darlington and York was cut by two minutes giving an average speed in excess of 80 mph; speeds in excess of 100 mph were regularly achieved. Without doubt, the 'Deltics' were an asset on the East Coast main line.

Nothing lasts for ever and that is especially so on the railways. In many respects the 'Deltics' had sown the seeds of their own destruction. Upgrading of the track had been required to allow their capabilities to be fully utilized, whilst better rolling-stock was developed for the enhanced services they introduced. A growth in traffic on the routes could be attributed at least in part to the 'Deltics', but airlines were offering increasing competition and the introduction of the high-speed 'Inter-City 125s' heralded the coming end. By 1980, an increasingly standardized BR mainte-nance system was not geared to dealing with a separate class of only 22 locomotives, and old age was also a factor in the decision.

The final frantic months of service saw 'Deltics' on service trains across the Pennines to Liverpool and elsewhere, but it was the 'Farewell Specials' which took them to places unfamiliar, including Devon. Enthusiast hysteria reached a peak on January 2 1982 with the final run on BR tracks. Without a doubt, these locomotives were special; it was not merely the power or the noise or the shape, all of which attracted some enthusiasts, but there was that indefinable 'something' about them which set the 'Deltics' apart. As far as diesels are concerned, they may certainly be described as classics. No fewer than six of the 22 locomotives have been preserved, with the prototype *Deltic* having a home in the Science Museum.

Side view of 'Deltic' No 55 019 Royal Highland Fusilier.

BR DIESEL-HYDRAULIC 'WESTERN' CLASS '52'

Designer BR Western Region. **Introduced** 1961. **Wheel diameter** 3 ft 9 in. **Axle arrangement** C-C **Engines** Two, Maybach MD655. **Power** 2 × 1,007 kW (1,350 hp) at 1500 rpm. **Type of drive** Hydraulic, Voith L630rV. **Tractive effort** 66,770 lbs (maximum), 45,200 lbs (continuous at 14.5 mph). **Maximum speed** 90 mph. **Full weight** 108 tons. **Number built** 74.

The British Railways modernization plan of 1955 envisaged a replacement of steam by diesel traction. Slow-speed diesels with electric transmission were favoured by four of the regions but the Western stood alone and decided upon high-speed engines with hydraulic power transmission. It was not a case of the former Great Western Railway being individualistic to the end, for there were logical reasons behind the choice. Two small high-speed engines would occupy the same space as a single low-speed unit and would offer back-up in the event of a breakdown. In addition, they would be lighter and easier to remove for maintenance. The hydraulic form of transmission was favoured because the higher management considered that maintenance personnel could adapt to it more readily than they could to the electrical form. Unlike other British railways, the GWR had no experience with electric traction and so had few trained people.

A number of diesel-hydraulic designs were approved and constructed, the final being the high-powered 'Western', or 'D1000' Class in the original BR scheme. Diesel-hydraulic locomotives had been successfully operated in West Germany and it was to that country that the Western Region's managers turned for assistance with its designs, especially with regard to the 'Westerns'. Maybach engines were chosen because of the German experience and the fact that they were used for some of the earlier designs. The Voith hydraulic transmission had also been used for a number of previous designs though not in combination with Maybach engines. The bogies were of the C-C form which allowed for a high tractive effort, while the axle loading was comparatively low because of the small lightweight engines. This high starting tractive effort made the 'Westerns' suitable for hauling the heavy stone trains from the Merehead quarries.

The 'Westerns' were constructed at Swindon and Crewe, but both experienced delays due to delivery of parts. This was

Above Western Courier *as preserved in maroon livery.*

Below *Preserved 'Western' diesel* Western Ranger *with a service train entering Arley on the Severn Valley Railway.*

particularly the case with the transmission units ordered from Germany, and the result was that a number of otherwise complete locomotives had to be stored pending delivery of the units.

In service they performed up to expectation working both passenger and freight trains, but defects soon manifested themselves resulting in a serious reduction in availability. The bogies caused concern due to a lack of lateral flexibility which became evident at speeds above 80 mph. The relatively rigid arrangement was even considered dangerous and modifications were put in hand. Transmission shaft roller bearings seized on the initial member of the class with the result that all the 'Westerns' were temporarily withdrawn for inspection. Few faults resulted from the engines or transmission, but failure of other items gave the class a bad name, at least initially.

By the time all the 'teething problems' had been cured and the 'Westerns' were carving a name for themselves, the order went out that all diesel-hydraulic classes should be withdrawn. Reduction in traffic and the spread of main-line electrification on the London Midland Region had released many diesel-electrics and the non-standard hydraulics became dispensible. During those final years of operation in the early 1970s, the 'Westerns' like the 'Deltics', became something of a 'cult' with enthusiasts desperate for seats on service and special trains. Such was the enthusiasm that no fewer than seven examples were subsequently preserved, and that only six years after the demise of main-line steam.

By being different, the hydraulics ensured a place for themselves in British locomotive history and the 'Westerns' particularly so because of their power and performance. Though many steam enthusiasts consider diesels to be 'just boxes', there was something special about a 'Western's shape. It had style and could even be considered as attractive. Brunel once told a locomotive supplier that Stephenson's *North Star*, the first GWR locomotive of any merit, 'would be the most beautiful ornament in the most elegant drawing-room'. The same might almost be said of the 'Westerns'.

BR DIESEL-ELECTRIC CLASS '47'

Designer Brush/BR. **Introduced** 1962. **Driven wheel diameter** 3 ft 9 in.
Axle arrangement Co-Co. **Engines** Sulzer 12LDA28C. **Power** 2,050 kW
(2,750 hp) at 800 rpm (later downgraded to 2,580 hp). **Type of drive**
Electric, six Brush TM 64.68 axle-hung motors. **Tractive effort** 62,000 lbs
(maximum), 30,000 lbs (continuous at 27 mph). **Maximum speed** 95
mph. **Full weight** 110-123 tons depending on type. **Number built** 542.

The original British Railways dieselization Pilot Scheme produced
two Type '4' classes, the Class '40s' and the 'peaks'. Neither
completely satisfied the Eastern Region which considered that a
Co-Co locomotive of similar power would be more suitable. Brush
decided that there was some merit in the idea, particularly as the
Sulzer LD engine fitted to the 'peaks' had been up-rated to 2,750
hp.

The design of the locomotive turned out to be quite radical in
that the machinery was housed in a lightweight 'monocoque' body
of stressed skin construction (this form of construction basically
meant that a heavy conventional chassis was not required). Cast
manganese steel commonwealth bogies were used which,
although heavy, were of a well-tried and proven design. BR did,
however, stipulate that the cabs were to be flat-fronted and electric
train heating equipment must be provided on the initial batch of
twenty locomotives. A feature of the agreement reached with
Brush was that BR could later build the locomotives under license
using Brush-manufactured equipment, and this they did at Crewe
works over a period of many years.

The initial batch of twenty performed well on the Eastern Region
for all secondary duties and even as stand-ins for failed 'Deltics'.
Quantity production was soon sanctioned, and trials on the
Western Region showed that they were capable of sustained
speeds in excess of 100 mph, so the original permitted maximum
was increased from 90 mph to 95 mph. Freight trains of 1,500 tons
were regularly hauled but this resulted in certain engine problems,
the outcome of which was that the engines were permanently
downgraded to 2,580 hp.

Large-scale production by Brush and BR at Crewe resulted in an
extensive fleet which served in many parts of Britain. Class '47s'
were, and are, used for all types of train from express passenger to
freight and may be considered as the diesel equivalent of that
classic mixed-traffic steam locomotive, Stanier's 'Black Five'.
Minor changes in detail do occur throughout the class but the main

Class '47' No 47 549 backs on to its train at Liverpool Lime Street in August 1982.

details as given above generally apply. Not all engines were provided with train heating boilers, some being specifically designated for freight working, and as steam heating has gradually been phased out, electric train heating systems have been installed.

Twelve Class '47s' have been fitted with equipment to allow for 'push-pull' operation on the Glasgow-Edinburgh high-speed services, while haulage of 'merry-go-round' trains supplying coal to power stations has required some members of the class to be provided with the Brush-developed automatic slow speed control system. This is essential to enable the locomotive to haul its train of loaded coal hoppers through the unloading plant at a speed of 0.5 mph.

It may seem strange that the '47s' have been included in a list of classic locomotives, especially as it is a diesel design which is still in regular widespread service. That is, in fact, the reason. They are reliable and have a power capability to suit most modern requirements, and more than 25 years after their first introduction the class is still an important part of the BR locomotive fleet, and is likely to remain so for the foreseeable future.

GLOSSARY

Tractive effort This is a rather vague concept of a locomotive's ability to haul loads. For steam locomotives it is derived by means of an empirical equation, Phillipson's formula, which makes use of certain critical dimensions and 85% of the boiler pressure:

$$\text{Tractive effort (lbs)} = \frac{0.85 \, d^2 \, s \, n \, p}{2w}$$

where d = cylinder diameter (inches)
     s = piston stroke (inches)
     n = number of cylinders
     p = boiler pressure (psi)
     w = driving wheel diameter (inches)

For diesel locomotives, the continuous and maximum tractive efforts are given, these being more accurate indications as to true hauling power.

Valve gear This is the means whereby the valves which direct steam to and from the cylinders may be actuated.
Caprotti – a system of poppet valves operated by a rotating shaft.
Conjugated – an arrangement whereby the inside cylinder valve gear on a three-cylinder engine is operated by the combined motions of the outside cylinder valve shafts.
Stephenson – A common arrangement using two eccentrics driven by the crankshaft.
Walschaerts – An arrangement making use of the combined motion effect of a crankshaft-driven eccentric and the piston crosshead.

Wheel arrangements For steam locomotives, the number of wheels provides the designation of type. The first number in the group of three represents the wheels on the leading bogie or pony truck. The second number indicates the wheels actually driven by the cylinders, whilst the third gives the number of wheels on the trailing pony truck or bogie. For example, '4-6-2' indicates four wheels on the leading bogie, six driven by the cylinders and two on the trailing pony truck. Leading wheels are used to guide a locomotive into a curve, thus reducing the flange wear on the driving wheels. Trailing wheels provide support for the rear part of the locomotive and are of use where a wide firebox prevents the driving wheels from being fitted close to the cab. Common named wheel arrangements are 'Atlantic' (4-4-2), 'Mogul' (2-6-0) and 'Pacific' (4-6-2).

For diesels, letters are used to indicate the number of powered axles on a bogie. 'A' indicates one powered axle, 'B' indicates two and 'C' three. The letter 'o' after the main letter means that each axle has its own power drive. A number before the main letter indicates the number of non-powered axles. For example, '1Co-Co1' means two bogies each with three powered axles and a non-powered axle. A 'C-C' arrangement means two bogies each with three axles but with only one common power drive on each bogie.

'Deltic' No 55017, The Durham Light Infantry, *at York.*